The

Pole Shift

That

Sank Atlantis

Message from the Ancients

The Pole Shift
That
Sank Atlantis

Message from the Ancients

By John Gagnon and Derek Cromie

Victor and Company

ISBN: 978-0-9811281-1-5

victorandcompanypublishing@gmail.com

Introduction

This book tells a different story than that of Mankind's recorded history. It challenges existing theories on such things as the Ice Ages, the related extinction of many plant and animal species and drastic changes in climate conditions. It offers a new understanding of who we are and where we came from.

A major catastrophic event took place only 12,400 years ago. Multiple large meteors impacted the Earth with such unimaginable force as to cause a major shift of the Earth's Axis. These findings have been corroborated by one of history's most important monuments, The Great Pyramid of Giza.

You will find compelling evidence that highly advanced civilizations existed before this event, civilizations that until now have always been considered a myth. The life of Mankind was changed forever.

Acknowledgements

We would like to thank our families for their support during the researching and writing of this book. Thank you from the bottom of our hearts.

Ten important works related to our topic are prominently featured in the book:

The Orion Mystery by Robert Bauval and Adrian Gilbert

Earth in Upheaval by Immanuel Velikovsky

Path of the Pole by Charles Hapgood

The Mystery of the Sphinx (video) by John Anthony West and Robert Schoch

Cataclysm by D.S. Allan and J.B. Delair

The Cycle of Cosmic Catastrophes by Richard Firestone, Allen West and Simon Warwick-Smith

Timeus by Plato

The Golden Age Project (website-www.goldenageproject.org.uk) by Edmund Marriage

Fingerprints of the Gods by Graham Hancock

Ocean Floor Map by Marie Tharp and Bruce Heezen

We have highlighted these researchers in hopes that their groundbreaking insights will gain ever-greater recognition. The information contained in their works afforded us the opportunity to consider all the possibilities and we greatly appreciate their efforts. We would like to personally give credit to these independent authors, many of whom suffered through dramatic criticisms while bringing forward their concepts and sharing their insights with us. Tally-Ho.

Contents

Chapter 1

Cataclysm

A Recent Extinction Event

It is a fact that there was a vast extinction of plants and animals near the end of the time period history refers to as the "Ice Ages". This most recent so-called Ice Age ended suddenly only about 12,000 years ago. Before this there were an astonishing number of odd and unique creatures that roamed this planet *alongside modern man* including over twenty species of elephant. Now there are only two elephant species, the African and the Asian.

From "Earth in Upheaval"

Many forms of life, many species and genera of animals that lived on this planet in a recent geological period, in the age of man, have utterly disappeared without leaving a single survivor. Mammals walked in fields and forests, propagated and multiplied, and then without a sign of degeneration vanished.

A considerable group have become extinct virtually within the last few thousand years. The large mammals that died out [in America] include all the camels, all the horses, all the ground sloths, two genera of musk-oxen, peccaries, certain antelopes, a giant bison with a horn spread of six feet, a giant beaverlike animal, a stag-moose, and several kinds of cats, some of which were of lion size. Also the Imperial elephant and the Colombian mammoth, animals larger than the African elephant and common all over North America, disappeared. The mastodon that inhabited the forests and ranged from Alaska to the Atlantic coast and Mexico, and the woolly mammoth that roamed in a broad area adjacent to the ice sheets, likewise persisted until a few thousand years ago.

The dire wolf, the saber-toothed tiger, the short-faced bear, the small horse (Equus tau) disappeared, and are no longer found either in the Old or in the New World. Many birds, too, became extinct.

These species are believed to have been destroyed to the last specimen in the closing Ice Age. Animals, strong and vigorous, suddenly died out without leaving a survivor. The end came, not in the course of the struggle for existence-with the survival of the fittest. Fit and unfit, and mostly fit, old and young, with sharp teeth, with strong muscles, with fleet legs, with plenty of food around, all perished.

These facts, as I have already quoted, drive "the biologist to despair as he surveys the extinction of so many species and genera in the closing Pleistocene [Ice Age]."

Some possible reasons given in explanation of this extinction event are disease, climate change, and over-hunting by man. We should assume that the evolution of the number of species we are considering here would have taken millions of years, and each had proven capable of surviving and adapting on this planet. Yet within a very short time period, *over half* of all large mammal species disappeared forever. Man did not kill off these creatures; if anything it is more likely that man was equally hunted.

Figure 1 – The Saber-toothed Tiger of the Ice Ages (by Charles R. Knight)

Could it be that only 12,400 years ago (10,400 BC), in a sudden and cataclysmic event, all species were brought to the brink of extinction, including man? Did highly advanced civilizations vanish into the mists of myth?

What we have to realize is that 12,400 years ago a major catastrophic event took place. When considering such a distant time, it places this event in a position that is very difficult for us to truly comprehend. Events that occur today, of even small significance, are within 24 hours known worldwide. The survivors of this cataclysm had no option but to make their own conclusions as to the events that changed their civilizations forever. The remaining survivors were forced in a very dramatic fashion to immediately change their inherent practices in order to survive. In today's society, even with all our technologic advances, this would be a mammoth task. Indeed we might even be at a disadvantage, for the knowledge of how to sustain ourselves if left on our own is lost for most of us.

We can only imagine the radical changes that must have taken place and the great difficulties that the remaining societies must have endured to ensure their survival. The efforts put forward by these survivors warrant our efforts to fully understand this time in our history.

Other Extinction Events

This recent "Ice Age" extinction event was only one of many in Earth's history. The extinction of the Dinosaurs, some 65 million years ago, is the most widely recognized. For many years now, that extinction event has widely been accepted to have been caused by a meteor impact and science has told us the resulting crater has been found (the Chicxulub crater in Mexico).

Recent evidence though has raised doubt about this crater. A few scientists, led by Professor Gerta Keller of Princeton and Professor Wolfgang Stinnesbeck of the University of Karlsruhe, have put forth geological evidence that suggests this is not the "smoking gun". Core samples of the sediments around this crater appear to show this impact happened well before the Dinosaurs went extinct. Evidence still indicates that the Dinosaur extinction was caused by meteor impact just that the resultant crater has not been found yet.

The most important point and lesson that we should take from these observations is that over years, with the constant gathering of new information, the scenarios we accept as historical facts often change. We at one point argued vigorously that the Earth was flat. We must now open our minds and understand that the Theory of Mankind's Prehistory may be running in the second position behind the Flat Earth Theory. The hardest part of accepting a new theory is acknowledging the errors in the existing theory and to make matters even more complicated, when one theory is found to be incorrect, other existing theories may be challenged. We are creatures of habit and changes are not easily accepted. We must push ourselves to find the truth and be willing to examine all possibilities. We must keep an open mind.

Let's turn our attention back to the Chicxulub crater. Half of it is buried beneath a landmass, the Yucatan Peninsula in Mexico, and the other half is below the ocean waters of the Gulf of Mexico. It is thought the crater was originally over 5 miles deep and well over 100 miles wide yet there is almost no visible sign of this crater on the surface of the Yucatan Peninsula. So how is it that this landmass came to be situated over top of just one half of the crater? Is it possible the crater, originally on the ocean floor, filled in with sediments over a long period of time and then suddenly the Yucatan Peninsula shifted over top of it? This type of scenario could well cause confusion for Science and will be discussed further in Chapter 7.

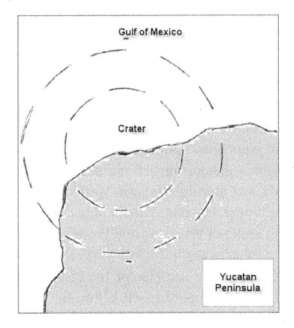

Figure 2 - Chicxulub Crater below Yucatan landmass

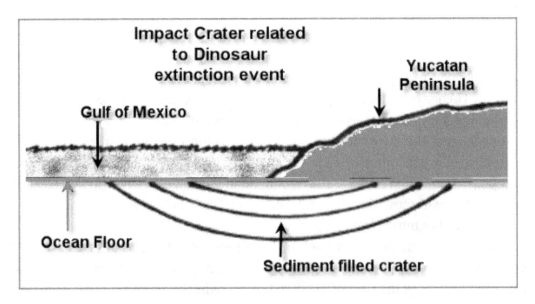

Figure 3 - Chicxulub Crater profile

More distant in time (over 200 million years ago) another major extinction event took place, the Permian-Triassic Extinction. The cause of this mass extinction has long been debated, but scientific attention has recently focused on a newly discovered impact crater that dates to this time period. The crater, found near Antarctica, is the result of the impact of a massive meteor that may have caused the greatest extinction event the world has ever known. It is thought that well over 90% of all plant and animal species on Earth went extinct.

We now know of two devastating extinction events that have been linked to separate impact events. Is it possible that most, if not all, large extinction events are related to large impact events? Do we have any other explanations for these mass extinctions? None as realistic as this! *Was this latest "Ice Age" extinction event also a result of meteor impact? The ancient Egyptians say it was so.*

Ivory in Siberia

One of the many unsolved mysteries pertaining to the end of the so called "Ice Ages" was the discovery of vast amounts of animal remains related to this time period that have been preserved (frozen) in a non-decomposing state, in areas foreign to their natural habitat. A never-ending supply of Mammoth tusks has turned up in Siberia, one of the coldest and most inhospitable places on Earth.

From "Path of the Pole"

Fossil tusks of the mammoth-an extinct elephant-were found in northern Siberia and brought southward to markets at a very early time, possibly in the days of Pliny in the first century of the present era. The Chinese excelled in working delicate designs in the ivory, much of which they obtained from the north. And from the days of the conquest of Siberia (1582) by the Cossack Yermak under Ivan the Terrible, until our own times, trade in mammoth tusks had gone on. Northern Siberia provided more than half the world's supply of ivory, many piano keys and many billiard balls being made from the fossil tusks of these mammoths.

In 1797 the body of a mammoth, with flesh, skin, and hair, was found in northeastern Siberia, and since then bodies of other mammoths have been unearthed from the frozen ground in various parts of that region. The flesh had the appearance of freshly frozen beef; it was edible, and wolves and sledge dogs fed on it without harm.

The ground must have been frozen ever since the day of the entombment; had it not been frozen, the bodies of the mammoths would have putrefied in a single summer, but they remained unspoiled for some thousands of years. "It is therefore absolutely necessary to believe that the bodies were frozen up immediately after the animals died, and were never once thawed, until the day of their discovery."

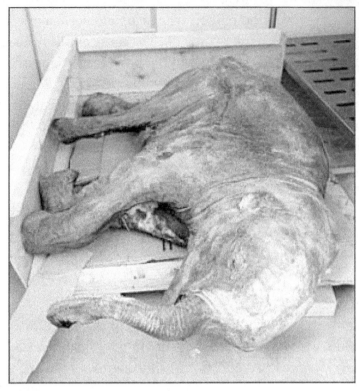

Figure 4 – 12,000 years old frozen baby Mammoth

How can we possibly explain the recent existence of large plant eating animals (perhaps numbering in the millions) discovered in an environment that today is so extreme that plant existence is almost impossible (never mind they supposedly thrived there during the "Ice Ages"). Even on islands in the Arctic Ocean north of Siberia, as recently as 200 years ago, were discovered the frozen remains of mammals not known to exist in such an extreme environment.

From "Earth in Upheaval"

The New Siberian Islands discovered in 1805 and 1806, as well as the islands of Stolbovoi and Belkov to the west, present the same picture. "The soil of these desolate islands is absolutely packed full of the bones of elephants and rhinoceroses in astonishing numbers." These islands were full of mammoth bones, and the quantity of tusks and teeth of elephants and rhinoceroses, found in the newly discovered island of New Siberia, was perfectly amazing, and surpassed anything which had as yet been discovered.

Even Charles Darwin stated that these facts were beyond his explanation and to this day no explanation is agreed upon.

From "Earth in Upheaval"

Georges Cuvier, the great French paleontologist (1769-1832), thought that in a vast catastrophe of continental dimensions the sea overwhelmed the land, the herds of mammoths perished, and in a second spasmodic movement the sea rushed away, leaving the carcasses behind. This catastrophe must have been accompanied by a precipitous drop in temperature; the frost seized the dead bodies and saved them from decomposition. In some mammoths, when discovered, even the eyeballs were still preserved.

Charles Darwin, who denied the occurrence of continental catastrophes in the past, in a letter to Sir Henry Howorth admitted that the extinction of mammoths in Siberia was for him an insoluble problem. J. D. Dana, the leading American geologist of the second half of the last century, wrote: "The encasing in ice of huge elephants, and the perfect preservation of the flesh, shows that the cold finally became suddenly extreme, as of a single winter's night, and knew no relenting afterward."

In the stomachs and between the teeth of the mammoths were found plants and grasses that do not grow now in northern Siberia. "The contents of the stomachs have been carefully examined; they showed the undigested food, leaves of trees now found in Southern Siberia, but a long way from the existing deposits of ivory. Microscopic examination of the skin showed red blood corpuscles, which was a proof not only of a sudden death, but that the death was due to suffocation either by gases or water, evidently the latter in this case. But the puzzle remained to account for the sudden freezing up of this large mass of flesh so as to preserve it for future ages."

Is it possible the explanation lies outside the bounds of currently accepted knowledge? Let us "think outside the box" and explore that possibility together. For example, could the explanation be that due to a cataclysmic impact event the entire landmass on which these animals lived was shifted a great distance in a northerly direction and the resulting drop in temperature was enough to preserve them to this day? *Alternatively,* did the North Pole shift suddenly to a position closer to Siberia (current pole), bringing with it a dramatic drop in temperature. As we shall see, The Great Pyramid does indicate that one of these two scenarios did happen!

Alaskan "Muck"

In Alaska, similar to Siberia but in more extreme concentrations, very large deposits of frozen animal remains have been found. These deposits contain both extinct and non-extinct animals and are thought to have been formed under catastrophic circumstances. Once again these animal parts have remained frozen continually for many thousands of years. We know that many of these animal species could not and would not survive in Alaska's present climate.

From "Earth in Upheaval"

> In Alaska, to the north of Mount McKinley, the tallest mountain in North America, the Tanana River joins the Yukon. From the Tanana Valley and the valleys of its tributaries gold is mined out of gravel and "muck." This muck is a frozen mass of animals and trees.
>
> F. Rainey of the University of Alaska described the scene: "Wide cuts, often several miles in length and sometimes as much as 140 feet in depth, are now being sluiced out along stream valleys in the Fairbanks District. In order to reach gold-bearing gravel beds an overburden of frozen silt or 'muck' is removed with hydraulic giants. This 'muck' contains enormous numbers of frozen bones of extinct animals such as the mammoth, mastodon, super-bison and horse. "These animals perished in rather recent times; present estimates place their extinction at the end of the Ice Age or in early post-glacial times. The soil of Alaska covered their bodies together with those of animals of species still surviving."
>
> Under what conditions did this great slaughter take place, in which millions upon millions of animals were torn limb from limb and mingled with uprooted trees? F. C. Hibben of the University of New Mexico writes: Although the formation of the deposits of muck is not clear, there is ample evidence that at least portions of this material were deposited under catastrophic conditions. Mammal remains are for the most part dismembered and disarticulated, even though some fragments yet retain, in their frozen state, portions of ligaments, skin, hair, and flesh. Twisted and torn trees are piled in splintered masses. ... At least four considerable layers of volcanic ash may be traced in these deposits, although they are extremely warped and distorted

What is of great interest are the conditions in which these remains are found. Very large numbers of dismembered animal parts are cemented together with twisted trees and boulders and other sediments.

In recent years we have seen the devastation that tsunamis and hurricanes can cause. Water has the power to move almost everything in its path. Eventually the resulting waves decrease in energy and recede leaving deposits of twisted wreckage. The Alaskan "Muck" deposits are similar, only on a much larger catastrophic scale. The formation of these deposits is thought by many to have occurred at the closing of the so-called "Ice Age" around 12,000 Years ago.

Bone Graves

In many locations around the world, of more moderate climate, similar evidence of animal remains has been found in rock formations such as cracks and caves. These locations hold the bones of numerous varied species (many of which are now extinct) completely fragmented and mixed together and covered with layers of sediment that enabled their preservation.

From "Earth in Upheaval"

On the Mediterranean coast of France there are numerous clefts in the rocks crammed to overflowing with animal bones. Marcel de Serres wrote in his survey of the Montagne de Pedemar in the Department of Gard: "It is within this limited area that the strange phenomenon has happened of the accumulation of a large quantity of bones of diverse animals in hollows or fissures." De Serres found the bones all broken into fragments, but neither gnawed nor rolled.

The Rock of Gibraltar is intersected by numerous crevices filled with bones. The bones are broken and splintered. "The remains of panther, lynx, caffir-cat, hyena, wolf, bear, rhinoceros, horse, wild boar, red deer, fallow deer, ibex, ox, hare, rabbit, have been found in these ossiferous fissures. The bones are most likely broken into thousands of fragments-none are worn or rolled, nor any of them gnawed, though so many carnivores then lived on the rock," says Prestwich, adding: "A great and common danger, such as a great flood, alone could have driven together the animals of the plains and of the crags and caves."

Scientists have shown that in locations where these bone graves have been studied, some of the species were completely foreign to the area in which they were found.

From "Earth in Upheaval"

"As Weidenreich began his studies, other amazing, nearly inexplainable features appeared." The fractured bones of seven human individuals were found there. "A European, a Melanesian, and an Eskimo type lying dead in one close-knit group in a cave, on a Chinese hillside! Weidenreich marveled." . . . It was assumed that the seven inhabitants of the narrow fissure were murdered because their skulls and bones are fractured. It is possible that these several types of man came together in Choukoutien, since the migrations of ancient man were on a greater scale than is generally thought.

But the finders of the conglomerates of bones were perplexed also by the animal remains: the bones belonged to animals of the tundra, or a cold-wet climate; of steppes and prairies, or dry climate; and of jungles, or warm-moist climate, "in a strange mixture." Mammoths and buffaloes and ostriches and arctic animals left their teeth, horns, claws, and bones in one great melange, and though we have met very similar situations in various places in other parts of the world, the geologists of China regarded their find as enigmatic.

Considering all previous information, was there a recent catastrophic event of a global nature with giant tsunamis and tidal surges repeatedly pounding every coast of every continent?

Whales in the Mountains

One of the most crucial pieces of evidence is the discovery of whalebones as high as 600 feet above sea level. Perhaps you have been lucky enough to see whales in their natural environment as well as salmon in the wild driven by Mother Nature's call for continuation of the species. On the salmons' journey upstream to mate they regularly jump rock cuts as high as 6 feet, but in all those wonderful years of observing Mother Nature in its natural state have you ever seen those special Whales that are capable of jumping 600 feet?

From "Earth in Upheaval"

In bogs covering glacial deposits in Michigan, skeletons of two whales were discovered. Whales are marine animals. How did they come to Michigan in the post-glacial epoch? Whales do not travel by land. Glaciers do not carry whales, and the ice sheet would not have brought them to the middle of a continent. Besides, the whale bones were found in post-glacial deposits. Was there a sea in Michigan after the glacial epoch, only a few thousand years ago?

In order to account for whales in Michigan, it was conjectured that in the post-glacial epoch the Great Lakes were part of an arm of the sea. At present the surface of Lake Michigan is 582 feet above sea level. Bones of whale have been found 440 feet above sea level, north of Lake Ontario; a skeleton of another whale was discovered in Vermont, more than 500 feet above sea level; and still another in the Montreal-Quebec area, about 600 feet above sea level.

The theory put forward by science for the existence of whale bones found high above sea level is that during the Ice Ages this land was depressed over 1,000 feet by weight of ice, allowing creation of a vast inland sea (the Champlain Sea). The theory goes that this land has since rebounded upward due to the ice's disappearance. This theory is questionable at best as these areas are located at the very edge of current rebound (areas of very minimal rebound) and areas just south of where these bones have been found are actually sinking today, not rising.

Another concern is that the Champlain Sea didn't even reach Michigan. It is surprising how often science simply ignores such inconvenient truths.

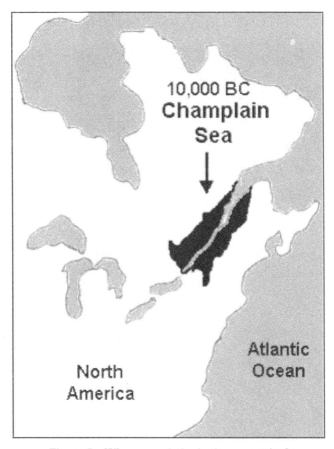

Figure 5 – Why were whales in the mountains?

Whalebones were not only found in small numbers at great height, they were also found in great quantities in locations far inland.

From "Earth in Upheaval"

A species of Tertiary whale left its bones in great numbers in Alabama and other Gulf States. The bones of these creatures covered the fields in such abundance and were "so much of a nuisance on the top of the ground that the farmers piled them up to make fences."

This is further evidence that the catastrophic events that took place in our ancient past were in large part water related. To what degree these extreme water conditions were caused directly by meteor impacts or the resultant planetary changes is something that needs further investigation.

Figure 6 – Just how advanced were the Ancients?

Chapter 2

Earth in Upheaval

Myth and Legend

Throughout history Mankind has handed down, through time, their individual stories of this catastrophic event that ended the Ice Age.

From "Cataclysm"

Yet the notion of a vast flood evidently is so ingrained in human memory that it could not be forgotten. It must, therefore, have been an event either of the most recent past or of exceptional proportions and severity.

An important feature in many of these traditions is that the cause of these calamities was one or more cosmic bodies which in simple terms engaged in a 'war in heaven'. That such a seemingly preposterous idea is common in so many different legends raises the question as to why numerous often-unrelated peoples should even share such a remembrance when no such celestial 'war' has been observed in historical times. The only conclusion that can be arrived at is that the disaster described in these traditions did actually occur, and made such an indelible impression upon the few survivors of the event that accounts of it were carefully preserved down to our own times as the most enduring of memories.

This world-changing event was so important to all of mankind that nearly every tribe and nation has to this day preserved this history through stories and legends.

From "Cataclysm"

Many catastrophe traditions refer to the coming of unnaturally prolonged darkness. The Central American Aztecs preserved such an account:

The third sun is called Quia-Tonatiuh, sun of rain, because there fell a rain of fire; all which existed burned; and there fell a rain of gravel. Now, this was in the year Ce Tecpalt, One Flint, it was the day Nahui-Quiahuitl, Fourth Rain. Now in this day, in which men were lost and destroyed in a rain of fire, they were transformed into goslings; the sun itself was on fire, and everything, together with the houses was consumed. A tremendous hurricane carried away trees, mounds, houses and the largest edifices, notwithstanding which many men and women escaped, principally in caves and places where the great hurricane could not reach them. All this time they were in darkness, without seeing the light of the sun, nor the moon, that the wind had brought them.

These myths and legends talk of flood and fire, celestial battles, the sky falling, prolonged darkness and daylight, colossal land movements and the sea boiling. For thousands of years, the elders have handed down such histories throughout the world.

In 1950 Immanuel Velikovsky released his book *Worlds in Collision* that had challenging theories for its time. His advanced theories were well written and researched and made excellent reading to a large number of people who helped this book achieve great success, becoming a best seller. His intuition was that ancient myths from around the world about cataclysmic events of a cosmic nature were perhaps not so mythical and this made a lot of sense to a lot of people. What's ironic is that the scientists of the day preferred to attack his theories rather than discuss them, even trying to have his book banned from learning institutions. Things are much the same today.

Mountains from Seafloors

The formation of mountain ranges has been an ongoing debate between scholars for more than a century. Most have argued that mountain ranges are formed in a very slow and direct movement. There are others who argue these mountain ranges were created in a much more accelerated time frame.

From "Earth in Upheaval"

To the surprise of many scientists, it was found that mountains have traveled, since older formations have been pushed over on top of younger ones. Chief Mountain in Montana is a massif standing several thousand feet above the Great Plains. It has been thrust bodily upon the much younger strata of the Great Plains, and then driven over them eastward, for a distance of at least eight miles. "Indeed, the thrust may have been several times eight miles", writes Daly.

"By similar thrusting, the whole Rocky Mountain Front, for hundreds of miles, has been pushed up and then out, many miles over the plains." Such titanic displacements of mountains have been found in many places on the earth. The displacement of the Alps is especially extensive.

"During the building of the Alps gigantic slabs of rock, thousands of feet thick, hundreds of miles long, and tens of miles wide, were thrust up and then over, relatively to the rocks beneath. The direction of the relative over thrusting movement was from Africa toward the main mass of Europe on the north. The visible rocks of the northern Alps of Switzerland have thus been shoved northward distances of the order of 100 miles. In a sense the Alps used to be on the present site of northern Italy." Mount Blanc was moved from its place and the Matterhorn was overturned.

Those portions of the Alps that surround the valley of the Linth, in the canton of Glarus in Switzerland, have lower parts of Tertiary formations or of the age of mammals; their upper parts are Permian (preceding the age of reptiles) and Jurassic (of the age of reptiles).

In the Alps, caverns with human artifacts of stone and bone dating from the Pleistocene (Ice Age) have been found at remarkably high altitudes. During the Ice Age the slopes and valleys of the Alps, more than other parts of the continent, must have been covered by glaciers; today in central Europe there are great glaciers only in the Alps. The presence of men at high altitudes during the Pleistocene or Paleolithic (rude stone) Age seems baffling. Could it be that the mountains rose as late as in the age of man and carried up with them the caverns of early man? In recent years evidence has grown rapidly to show, in contrast to previous opinion, that the Alps and other mountains rose and attained their present heights, and also traveled long distances, in the age of man.

Many mountains throughout the world have a remarkably youthful appearance. Their sharp jagged peaks are yet to be worn smooth by Mother Nature. Common erosion over millions of years will eventually convert a large mountain to a small hill.

From "Earth in Upheaval"

The Himalayas, highest mountains in the world, rise like a thousand-mile-long wall north of India. This mountain wall stretches from Kashmir in the west to and beyond Bhutan in the east, with many of its peaks towering over 20,000 feet, and Mount Everest reaching 29,000 feet, or over five miles. The summits of these lofty massifs are capped by eternal snow in those regions of the heavens where eagles do not fly nor any other bird of the sky.

Scientists of the nineteenth century were dismayed to find that, as high as they climbed, the rocks of the massifs yielded skeletons of marine animals, fish that swim in the ocean, and shells of mollusks. This was evidence that the Himalayas had risen from beneath the sea. At some time in the past azure waters of the ocean streamed over Mount Everest, carrying fish, crabs, and mollusks. Marine animals looked down to where now we look up.

Is it possible that catastrophic land movements are responsible for the creation of some of these mountain ranges? Did they rise up in just one day?

Charles Darwin

Charles Darwin spent five years exploring the wilds of South America. His astute observations lead to the future development of his theories on evolution and natural selection. These theories proved to be very important for that time in history, but to him many questions remained unanswered.

He was astonished by the amount of changes to this southern continent that must have occurred in recent times. His confusion was caused by the unexplainable nature of these changes. With the extinction of many animal species there came a dramatic change in the land.

From "Path of the Pole"

Sir Archibald Geike summarized Darwin's findings thus:
On the west coast of South America, lines of raised terraces containing recent shells have been traced by Darwin as proofs of a great upheaval of that part of the globe in modern geologic time. The terraces are not quite horizontal but rise to the south. On the frontier of Bolivia they occur from 60 to 80 feet above the existing sea level, but nearer the higher mass of the Chilean Andes they are found at one thousand, and near Valparaiso at 1300 feet.

Even Charles Darwin himself, frustrated in his search for a reasonable explanation, stated that for these events to have occurred *"we must shake the entire framework of the globe"*.

From "Earth in Upheaval"

Charles Darwin, who had previously dropped his medical studies at Edinburgh University, upon his graduation in theology from Christ College, Cambridge, went in December 1831 as a naturalist on the ship Beagle, which sailed around the world on a five-year surveying expedition. Darwin had with him the newly published volume of Lyell's Principles of Geology that became his Bible. On this voyage he wrote his Journal, the second edition of which he dedicated to Lyell.
This round-the-world voyage was Darwin's only fieldwork experience in geology and paleontology, and he drew on it all his life long. He wrote later that these observations served as the "origin of all my views." His

observations were made in the Southern Hemisphere and more particularly in South America, a continent that had attracted the attention of naturalists since the exploration travels of Alexander von Humboldt (1799-1804). Darwin was impressed by the numerous assemblages of fossils of extinct animals, mostly of much greater size than species now living; these fossils spoke of a flourishing fauna that suddenly came to its end in a recent geological age. He wrote under January 9, 1834, in the Journal of his voyage:

"It is impossible to reflect on the changed state of the American continent without the deepest astonishment. Formerly it must have swarmed with great monsters: now we find mere pigmies, compared with the antecedent, allied races."

He proceeded thus: "The greater number, if not all, of these extinct quadrupeds lived at a late period, and were the contemporaries of most of the existing sea-shells. Since they lived, no very great change in the form of the land can have taken place. What, then, has exterminated so many species and whole genera? The mind at first is irresistibly hurried into the belief of some great catastrophe; but thus to destroy animals, both large and small, in Southern Patagonia, in Brazil, on the Cordillera of Peru, in North America up to Behring's [Bering's] Straits, *we must shake the entire framework of the globe.*"

These questions have confused not only Charles Darwin, but also much of Mankind for generations. It is very intriguing that the one statement that Charles Darwin expressed in total frustration is the one statement that may be closest to the answer.

Ocean Floor Mystery

So far we have found whalebones 600 feet above sea level and the fossil remains of sea creatures on top of the highest mountains. Now we shall discover beach sand over two miles deep on the ocean floor.

From "Earth in Upheaval"

In the fall of 1949, Professor M. Ewing of Columbia University published a report on an expedition to the Atlantic Ocean. Explorations were carried on especially in the region about the Mid-Atlantic Ridge, the mountainous chain that runs from north to south, following the general outlines of the ocean. The Ridge, as well as the ocean bottom to the west and to the east, disclosed to the expedition a series of facts that amount to "new scientific puzzles." One was the discovery of prehistoric beach sand . . . "brought up in one case from a depth of two and the other nearly three and one half miles, far from any place where beaches exist today." One of these sand deposits was found twelve hundred miles from land.

Sand is produced from rocks by the eroding action of sea waves pounding the coast, and by the action of rain and wind and the alternation of heat and cold. On the bottom of the ocean the temperature is constant; there are no currents; it is a region of motionless stillness. Mid-ocean bottoms are covered with ooze made up of silt so fine that its particles can be carried suspended in ocean water for a long time before they sink to the bottom, there to build sediment. The ooze contains skeletons of the minute animals, foraminifera that live in the upper waters of the ocean in vast numbers. But there should be no coarse sand on the mid-ocean floor, because sand is native to land areas and to the continental shelf.

This is another of the many facts discovered in recent times that show worldwide evidence of an event so violent and powerful that it changed the face of the planet, as the survivors had known it. Yet another clue that adds legitimacy to the possibility of great Earth changes was the discovery that the Atlantic Ocean sea floor had very little sediment cover.

From "Earth in Upheaval"

But there was another surprise in store for the expedition. The thickness of the sediment on the ocean bottom was measured by the well-developed method of sound echoes. An explosion is set off and the time it takes for the echo to return from the sediment on the floor of the ocean is compared with the time required for a second echo to return from the bottom of the sediment, or from the bedrock, basalt or granite. "These measurements clearly indicate thousands of feet of sediments on the foothills of the Ridge. Surprisingly, however, we have found that in the great flat basins on either side of the Ridge, this sediment appears to be less than 100 feet thick.

Actually, the echoes arrived almost simultaneously, and the most that could be attributed in such circumstances to the sediment was less than one hundred feet of thickness, or the margin of error. "Always it had been thought the sediment must be extremely thick, since it had been accumulating for countless ages. ... But on the level basins that flank the Mid-Atlantic Ridge our signals reflected from the bottom mud and from bedrock came back too close together to measure the time between them ... They show the sediment in the basins is less than 100 feet thick."

The absence of thick sediment on the level floor presents "another of many scientific riddles our expedition propounded."

It has since been shown that throughout much of the world's oceans, and at the bottom of deep-sea trenches, the expected amount of sediments does not exist.

Chapter 3

The Path of the Pole

Sudden End to the "Ice Age"

Current scientific theory states that as recently as 12,000 years ago much of the North American continent was covered by vast ice fields as much as five miles thick. It is thought this ice field covered most of Canada and continued deep into the United States and similar ice fields are believed to have covered most of northern Europe and Greenland. This has historically been referred to as the "Ice Age". The formation of such quantities of ice would have taken many thousands of years and recent core samples taken from Greenland's ice sheets appear to date back hundreds of thousands of years. Greenland's ice sheets comprise the only remaining ice in the Northern Hemisphere that can be dated to this time in history. The *theorized* North American and European ice sheets completely disappeared at the close of the Ice Ages.

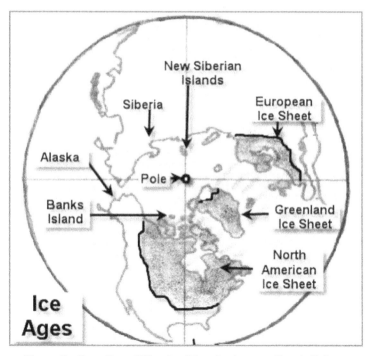

Figure 7 – Location of "Ice Age" ice sheets according to Science

For these massive ice fields to melt in such a short period of time there must have been a rapid and dramatic change in climate. To put this in perspective, we are told the amount of ice that melted was enough to raise the ocean levels by over 300 feet. Take a moment to consider this last point.

From "Path of the Pole"

> Our present objective is to establish, by marshaling all the necessary facts, the rate at which the Wisconsin (American) ice cap melted. If it can be shown, as I believe it can, that the ice sheet melted at a rate that is entirely inexplicable in the light of presently accepted theories of geology, we shall have established our right to look at new ideas. We shall have established that in this instance the concepts of uniformitarianism do not apply, that an exceptional cause must be found, but one that fundamentally reflects the dynamic realities of the strange planet on which we live.

A sudden repositioning of North America and Europe to warmer latitudes (to their current positioning) could account for this change in climate. Such a change would have left these existing ice fields in a permanent state of deterioration. Greenland would have been affected as well although it remained within the Arctic Circle, thereby preserving its ice sheets. *Evidence of rapid and dramatic climate change related to this time period has been found worldwide.*

From "Cataclysm"

> Essentially comparable changes also occurred about the same time much more distantly. We may highlight those from Tibet and China, where climate changes seriously affected Tibetan lake levels and led to long-term aridity across much of the Tibetan plateau . . .

Jorgen Peder Steffensen of the Niels Bohr Institute at the University of Copenhagen states "Our new and extremely detailed data from the examination of the ice cores (Greenland) shows that in the transition from the Ice Age to our current warm interglacial period, the climate shift was so sudden that it is as if a button was pressed." Did a giant meteor press this button? Did the dramatic climate changes of this time period result from a catastrophic impact event that repositioned the continents in relation to the poles?

The "Ice Age" Enigma

We are told by science that during the last Ice Age the polar ice cap covered most of North America, northern Europe and Greenland. A puzzling fact is that these land areas are all positioned on "one side" of today's pole. Just as important, although rarely discussed, is the fact that lands on the "other side" such as Alaska, Siberia, the New Siberian Islands, the northern tip of Greenland and islands north of Canada show no evidence of previous ice formations. All of these areas are presently located in a much closer proximity to the northern pole than are the formerly ice-bound regions.

From "Path of the Pole"

> Now we come to Banks Island, Northwest Territories, in Lat. 73° 21″ N, Long. 121° 54″ W, on which no evidence of glaciation has ever been found . . . Sir Robert McClure mentions the existence of a fossil forest on Banks island, where, for a depth of 40 feet, a cliff was composed of one mass of fossil tress … 120 miles further north we discovered a similar kind of fossil forest. As we shall see, such fossil forests have been found not only in the Canadian Arctic Archipelago but all along Siberia.

The facts indicate that there was a catastrophic Geo-repositioning of all landmasses at the close of the so-called Ice Ages. Areas previously glaciated were originally located closer to the Poles. Areas that show no sign of previous glaciations but are presently in closer proximity to the pole today must have been originally farther south. The primary evidence used to develop the "Ice Age" theory in the 1800s was the belief that ice had covered areas as far south as the USA and Europe. Therefore it was theorized that some vast ice cap must have existed in the past. The possibility of continental movement or pole shift was never a consideration at that time. Also, since the general acceptance of the "Ice Age" theory within academia, much of the evidence has been re-evaluated and quite frankly does not support this theory.

From "Cataclysm"

> Numerous lines of inquiry converge upon the startling fact that the Ice Age of orthodoxy is no more than the shaky theory it always has been and its former reality, as conceived by its advocates, just a wonderful myth.
>
> If, as demonstrated, the great ice sheets so beloved of glacialists never existed, because the uplands so necessary for their development and maintenance were either too low or non-existent during alleged Ice Age times, and because ice, even very thick ice, cannot behave in the manner required by glacial theory, it follows that other geological phenomena commonly ascribed to massive ice action were caused by some other agency or combination of circumstances. Among such phenomena may be mentioned moraines, striated rocks, rock flutings, giant surface grooves, cirques, hanging valleys, polished rock surfaces, kettle-holes, drumlins, eskers, kames and erratic boulders.

Ice at the Equator

Under no possible circumstances with our existing climate could an ice cap form in Africa. Yet the record reveals that not only Africa but also other locations near the equator show evidence that polar-like ice fields once existed there.

From "Earth in Upheaval"

> In 1865, Agassiz went to equatorial Brazil, one of the hottest places in the world, where he found all the signs he ascribed to the action of ice. Now even those who had previously agreed with him became distressed. An ice cover in the tropics, on the very equator? There were drift accumulations, and scratched rocks, and erratic boulders, and fluted valleys, and the smooth surface of tillite (rock formed of consolidated till), so there must have been ice to carry and polish, and the region must have gone through an ice period. What could have caused a tropical region to be covered by ice several thousand feet thick?
>
> Abundant vestiges of an ice age were likewise found in British Guiana, another of the hottest places on earth. Soon the same word came from equatorial Africa; and what appeared even more strange, the marks there indicated not only that equatorial Africa and Madagascar had been under a sheet of ice but that the ice had moved, spreading from the equator toward the higher latitudes of the Southern Hemisphere, or in the wrong direction. Then vestiges of an ice age were discovered in India, and there, too, the ice had moved from the equator, and not merely toward higher latitudes, but uphill, from the lowland up the foothills of the Himalayas.

The same evidence left behind by this so called "Equatorial Ice Age", which we are told occurred many millions of years ago, has also been left behind by this most recent "Ice Age", according to science. These facts indicate that similar events took place at different times in Earth's history. In both cases it appears that landmasses that are now in ice free areas were at one time located within an arctic region.

Earth Crust Displacement

Fifty years ago, in response to mounting evidence of recent Earth changes, Charles Hapgood developed the theory of the possible displacement of the Earth's crust. Upon organization and understanding of the scientific facts he proposed that just 12,000 years ago there occurred a dramatic movement of the Earth's crust (in its entirety) around its fluid core similar to that of an orange peel which is set free to move around the inner orange. Important to this theory is the fact that the Earth is not a perfect sphere. Its rotation around its axis has caused a bulge at the equator and flattened Polar Regions. Charles Hapgood theorized that the displacement of the Earth's crust over irregular surfaces might account for the formation of large cracks in the sea floor and mountain ranges.

This theory of Earth Crust Displacement (ECD) was developed by Mr. Hapgood in order to explain how the North Pole could have shifted from Hudson Bay in Canada. His extensive observations and investigations had indicated a strong possibility that the North Pole had been situated on North America in the recent past. *It is important at this point to understand that these are two separate theories (ECD and Hudson Bay Pole) and that even if one is found to be incorrect, the other should not be automatically dismissed.*

During this process of forming these profound new theories Mr. Hapgood contacted Albert Einstein to seek out his opinion. After reviewing Mr. Hapgood's materials, Mr. Einstein felt compelled to state that this theory certainly was credible and should be further explored. He was of such strong belief in fact that he agreed to write a Forward for Charles Hapgood's upcoming book "Path of the Pole". Here is that Forward

From "Path of the Pole"

I frequently receive communications from people who wish to consult me concerning their unpublished ideas. It goes without saying that these ideas are very seldom possessed of scientific validity. The very first communication, however, that I received from Mr. Hapgood electrified me. His idea is original, of great simplicity, and - if it continues to prove itself - of great importance to everything that is related to the history of the earth's surface.

A great many empirical data indicate that at each point on the earth's surface that has been carefully studied, many climatic changes have taken place, apparently quite suddenly. This, according to Hapgood, is explicable if the virtually rigid outer crust of the earth undergoes, from time to time, extensive displacement over the viscous, plastic, possibly fluid inner layers. Such displacements may take place as the consequence of comparatively slight forces exerted on the crust, derived from the earth's momentum of rotation, which in turn will tend to alter the axis of rotation of the earth's crust.

In a polar region there is continual deposition of ice, which is not symmetrically distributed about the pole. The earth's rotation acts on these unsymmetrically deposited masses, and produces centrifugal momentum that is transmitted to the rigid crust of the earth. The constantly increasing centrifugal momentum produced in this way will, when it has reached a certain point, produce a movement of the earth's crust over the rest of the earth's body, and this will displace the polar regions toward the equator.

Without a doubt the earth's crust is strong enough not to give way proportionately as the ice is deposited. The only doubtful assumption is that the earth's crust can be moved easily enough over the inner layers.

The author has not confined himself to a simple presentation of this idea. He has also set forth, cautiously and comprehensively, the extraordinarily rich material that supports his displacement theory. I think that this rather astonishing, even fascinating, idea deserves the serious attention of anyone who concerns himself with the theory of the earth's development.

"A recent Hudson Bay Pole" and "Displacement of the Earth's Crust" were important and interesting new theories. You will see that the secrets that were built into The Great Pyramid in combination with recent geological findings and other clues strongly suggest that Charles Hapgood was on the right track.

"Outside the Box"

Perhaps you will find that there is actually more fiction than fact surrounding this particular catastrophic event. The upcoming information in this book leads in one direction only; 12,400 years ago the Earth was struck by multiple large meteors, which resulted in profound changes to this planet and greatly altered Mankind's evolution. You may find the conclusions of the authors quoted to this point of interest.

From "Cataclysm"

This book is about the greatest single disaster known to have befallen Earth. The event was actually part of a still larger catastrophe which, commencing in interstellar space, eventually embraced most of the solar system. In terms of geological time it occurred extremely recently, although science has been decidedly reluctant to admit the fact, and modern humanity has largely forgotten it.

. . .

New mountain ranges and lofty plateaus had been upheaved, and deep gorges and huge fissures had come into being. Unimaginably large eruptions of magma, lava and other volcanic ejecta, including steam and gases, had attended these derangements. Former seas and lakes had been drained or drastically reduced - the displaced waters roaring as immense irresistible watery avalanches over the newly emerging landscape, as they sought new basins. Only the very largest obstacles withstood these churning torrents.

. . .

Clearly, members of several different races sometimes took refuge in the same caverns or on the same mountain peaks, being as often as not complete strangers to one another. Such mixed gatherings were not expected, for, faced with an all-encompassing common danger, the sole aim of every individual must surely have been survival. Previous social considerations such as race, rank, gender or age, became meaningless instantly.

From "Path of the Pole"

We have seen that the problem of the geographical stability of the poles has long been a vexatious matter for science. From time to time theories of polar shift have been advanced, supported by large quantities of evidence, but the proposed mechanisms have been found defective, and in consequence the theories have been rejected. The failure of the theories has led, in the following years, to neglect of the evidence or to its analysis in accordance with theories conforming to the doctrine of the permanence of the poles. Although all the older theories of polar change, including that of Wegener, have been discredited, the evidence in favor of polar change has constantly increased. As a consequence, many writers at the present time are discussing polar shift, but none of them has as yet suggested an acceptable mechanism.

. . .

Let me say a final word to the students, the young men and women in our colleges and high schools: The mysteries of the earth beckon to you. What man now knows is little enough, and most of his general concepts in every field are vitiated by the artificial concepts he has created to cover his ignorance. These concepts must be destroyed. One tool exists that can accomplish this destruction, and this tool is in your hands. It is simply curiosity—the instinct to ask and to question. It should be kept sharp and used without mercy.

From "Earth in Upheaval"

Thus from the geological evidence we came to the conclusion to which we had also arrived traveling the road of the historical and literary traditions of the peoples of the world—that the earth repeatedly went through cataclysmic events on a global scale, that the cause of these events was an extraterrestrial agent, and that some of these cosmic catastrophes took place only a few thousand years ago, in historical times.

Many world-wide phenomena, for each of which the cause is vainly sought, are explained by a single cause: The sudden changes of climate, transgression of the sea, vast volcanic and seismic activities, formation of ice cover, pluvial crises, emergence of mountains and their dislocation, rising and subsidence of coasts, tilting of lakes, sedimentation, fossilization, the provenience of tropical animals and plants in polar regions, conglomerates of fossils of animals of various latitudes and habitats, the extinction of species and genera, the appearance of new species, the reversal of the earth's magnetic field, and a score of other world-wide phenomena.

. . .

As important as the "world catastrophes" conclusion is, it grows in significance for almost every branch of science when, to the ensuing question, "Of old or of recent time?" the answer is given, "Of old and of recent." There were global catastrophes in prehuman times, in prehistoric times, and in historical times. We are descendants of survivors, themselves descendants of survivors. We read here a few pages from the logbook of the earth, a rock rolling in space, circling with its attendant lifeless satellite around a fire-breathing star, moving with this its primary and other revolving planets through the galaxy of the Milky Way of - hundreds of millions of burning stars, and together with this entire host, through the void of the universe.

Perhaps it is time for some accepted but unsubstantiated theories to be discarded. If we as Mankind are serious about solving this great mystery we must look at all the new evidence. Much of this hard evidence can no longer be ignored.

Perhaps now is the time to think outside this "box" that has been so readily accepted. It is now time for us as intelligent individuals to lead in this process rather than to be led.

Figure 8 – Have the Ancients left us a message?

Chapter 4

The Orion Mystery

Giza

On a plateau overlooking the Egyptian capital of Cairo, on the northeastern tip of Africa, lie the sacred grounds of Giza. Here lie some of the worlds most important ancient monuments, the two best known being the Sphinx and The Great Pyramid. Co-existing with these monuments are two other very large pyramids. All of these monuments have been constructed in a very precise manner.

Figure 9 - Giza

It has been thought that the pyramids of Giza were constructed about 4,500 years ago during the Fourth Dynasty of the ancient Egyptian civilization. Incredibly, this is only a few hundred years after ancient Egypt's founding, thought to be Mankind's first true civilization (You will see this date of construction was recently confirmed through star alignments built within the walls of The Great Pyramid). In the remaining 2,500 years of this Egyptian civilization such amazing feats of engineering were never duplicated.

How did such a relatively young Egyptian civilization obtain the knowledge to create such extremely large and precise monuments? Even today the methods of construction used remain a mystery. Is it not reasonable and possible this technology was handed down from the survivors of a lost, highly advanced civilization? It is hard to understand the principle of man hunting with a small stone to survive and shortly thereafter being capable of moving 70 ton stones to very precise locations high inside The Great Pyramid. It seems many pages of history remain to be written.

Within a short period of time, or by the beginnings of the Roman Empire, all detailed history concerning the structures at Giza had disappeared. It is therefore possible that information of advanced civilizations previous to the Egyptians was lost as well. There is a myth of the existence of a "Hall of Records", said to be located within the area known as Giza. Perhaps if found these records will answer many of these questions.

The Sphinx

Standing majestically in front of the pyramids is the great stone monument known as the Sphinx. Construction of the Sphinx would have been a huge undertaking; it was formed out of solid bedrock with the removal of hand cut slabs of stone weighing as much as 100 tons. These slabs were then moved large distances and stacked to build temples. This removal took place completely around the perimeter of the Sphinx creating a deep rectangular enclosure that left the center portion untouched. This portion was then carved into the Sphinx.

Mainstream Egyptian historians believe the construction of the Sphinx took place within the same historical period as the pyramids. John Anthony West and Robert Schoch have argued otherwise. In their video "The Mystery of the Sphinx" they established evidence that the Sphinx may have been constructed many thousands of years before the pyramids. One of the most profound points of evidence they introduced is the existence of long smooth channels of erosion found on both the Sphinx's body and the walls of its enclosure. This erosion is very typical of that caused by water over long periods of time. It is known that Giza has been within a desert area since long before the construction of its pyramids and therefore, accepting the fact of water erosion, the Sphinx must have been constructed at a far earlier time when Giza received significant rainfall.

Figure 10 - Water Erosion of Sphinx Enclosure

John Anthony West and Robert Schoch have shown that the Sphinx is likely much older than the ancient Egyptian civilization known to us. This strongly indicates that the builders of the Sphinx belonged to a highly advanced earlier culture whose existence is currently unknown.

The Great Pyramid

In discussing the three large pyramids at Giza there is one notable difference between them. Although all three are built to very high standards, The Great Pyramid was built to standards that would be extremely difficult to match today. Its base is perfectly level and aligned to the cardinal points. Over 2,300,000 stones ranging from two tons to seventy tons were used in its construction. It was originally faced with twenty ton blocks of gleaming white limestone, most of which are now missing, so that all four sides were perfectly smooth (these missing face stones were used to build Cairo in ancient and later times). Of the remaining face stones it has been stated "there is no place a knife can be inserted between them".

Figure 11 – Facing stones

The only known entrance built into The Great Pyramid was a stone swivel door that weighed over twenty tons, but could be opened from the inside using minimal force and when closed became invisible with no crack wide enough to pry it open. The Great Pyramid remained the tallest man made structure for over 4000 years.

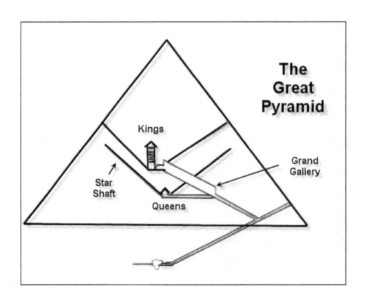

Figure 12 - Internal features of the Great Pyramid

Pyramids were normally built with chambers below ground level. The Great Pyramid though, which was built to more precise standards, contains one rough (unfinished) chamber below ground and three very distinct and exquisite chambers above. The upper chambers consist of the Kings Chamber, the Queens Chamber and the spectacular Grand Gallery.

Figure 13 – King's Chamber

Another very distinct difference in the internal structure of The Great Pyramid is that it is the only pyramid that contains so called "shafts", the purpose of which is still disputed. Two of these shafts extend from the Kings Chamber and two from the Queens Chamber. Each of the four shafts extend upward towards the surface of the pyramid although with some crucial differences, each shaft has been constructed at a different angle and the shafts of the Kings Chamber were open at each end while the shafts of the Queens Chamber do not reach the outside of the pyramid and originally stopped just short of the interior walls of the chamber (they were only revealed in recent times when investigation of a hollow sound behind a wall in the Queens Chamber led to their discovery). Why the original construction called for the complete entombment of the two shafts of the Queens Chamber has perplexed many.

Figure 14 – The Great Pyramid

The Great Pyramid is so massive that the temperature of the interior chambers remains constant at 20°C, the average temperature of the Earth. At this point we must all agree that The Great Pyramid was built with great effort and precision. Why was so much care taken in its construction? The builders must have felt that its precise construction and location was of extreme importance.

"As Above, So Below"

Throughout modern history it has always been a mystery why the two larger pyramids at Giza are in perfect alignment and are similar in size, while the third pyramid is smaller and has been purposely located in an offset position. Robert Bauval's theory that the three pyramids represent the three stars of Orion's Belt was for him and Mankind a very important discovery. He realized that Orion's Belt, located in the constellation Orion, consisted of two larger very bright stars and a third star that visibly appears to be smaller and located in an offset position. The long held myth that Egypt was "Heaven's Mirror" now seemed possible. Mr. Bauval then realized that other pyramids had been built outside Giza in locations that mirrored the positions of other stars that were part of the Orion constellation. It also appeared the location and direction of the Nile river mirrored the Milky Way, the long densely packed band of stars that runs through the night sky (which is our view of our home Galaxy seen edge-on).

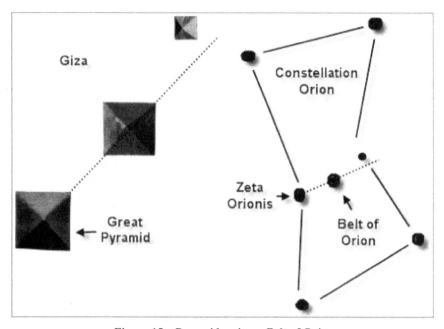

Figure 15 – Pyramids mirror Belt of Orion

Robert continued his investigation. With his knowledge of the shafts within The Great Pyramid (long 8 inch by 8 inch tunnels), he wondered if one of them might have been positioned to target the star in Orion's Belt that The Great Pyramid mirrored; that is, the star Zeta Orionis. With the aid of Skyglobe, a computer program that can track star positions far back in time, he was amazed and reassured to find that his theory of corroboration between Orion's Belt and the pyramids of Giza was accurate.

In 2,450 BC, very close to the generally accepted date of construction for The Great Pyramid, the star Zeta Orionis in Orion's Belt would have been in perfect alignment with the southern star shaft of the Kings Chamber. Bauval investigated further and realized that at the same time (2,450 BC) the northern shaft of the Kings Chamber pointed directly at Alpha Draconis in the constellation Draco. This was another very definitive star, being the northern

Pole Star of that time. These shafts were indeed "star shafts" and this discovery strongly validated the pyramids actual time of construction.

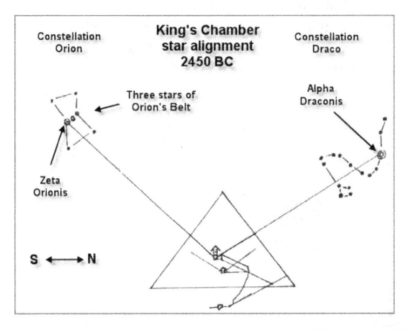

Figure 16 - King's Chamber Star Alignment 2450BC

The reason Robert Bauval had to use Skyglobe was that over thousands of years each star and constellation does slowly change position. These changes are barely noticeable over a lifetime yet you will see the ancient Egyptians understood them well. As a result of this slow movement, star positioning is a great marker of "time and place". This is the key to The Great Pyramid.

Observations of the stars reveal that every night the stars slowly travel across the night sky from east to west just as the sun does. This illusion is caused by the rotation of the Earth around its axis; it is the Earth that is moving, not the stars. From a point on Earth such as Giza you can divide the stars into two groups, those you see looking south and those you see looking north. If you pick the spot on the horizon that is directly south (same goes for north) and imagine a line from that point going up to directly above you, then each star in the southern sky would cross this line each night. If the point on the horizon is zero degrees (0°) and vertically overhead is 90°, then the angle that a star is at as it crosses this "Meridian" line is the angle that the star shafts in The Great Pyramid represent. Below we see representations of Orion crossing the Southern meridian at Giza in 2,450 BC.

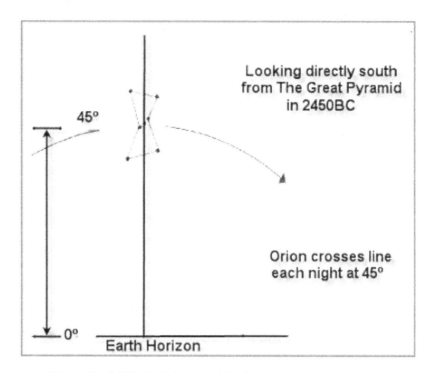

Figure 17 – 2450BC: Orion "meridian" crossing as seen from Giza

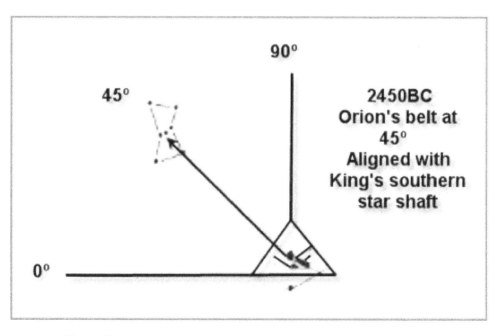

Figure 18 - Star Shaft Alignment with Orion meridian crossing - 2450BC

The King's Chamber star shafts were constructed such that each night in 2,450BC, not only could the meridian crossing of Zeta Orionis (in Belt of Orion) be "sighted" through the southern shaft, but also the meridian crossing of Alpha Draconis (the northern Pole Star) could be "sighted" through the northern shaft. The next diagram shows how the angles of the Queen's Chamber shafts differ from those of the King's Chamber. Each shaft has been meticulously constructed at a particular angle to precisely mark the Meridian crossing of a star.

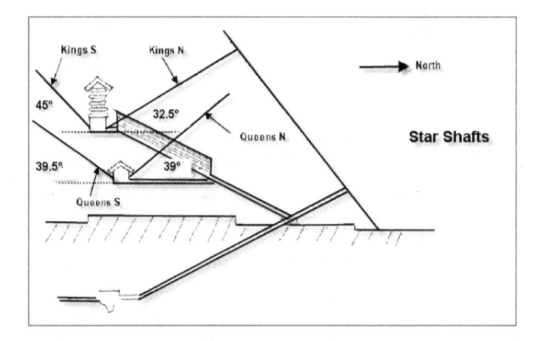

Figure 19 – Star shaft angles

In regard to the Queen's shafts, Robert Bauval and others have offered the tracking of stars other than Zeta Orionis and Alpha Draconis (the four shafts mark four different stars). We hope to show that the King and Queen's star shafts both reference the same two stars, but not in any way that has previously been proposed, each chambers two star shafts mark the same two stars (Zeta Orionis and Alpha Draconis) but at vastly different times.

How the Earth Wobbles (Precession)

As the Earth circles the sun its axis continually points at the same two stars (North and South Pole stars). The angle of Earth's axis, in relation to the plane of Earth's orbit around the sun, is called the Ecliptic. Earths Ecliptic stays fairly constant around 23° (varies between 21.5 to 24.5 degrees).

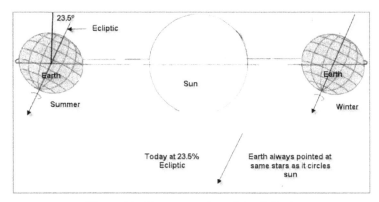

Figure 20 - Representation of the Ecliptic

More important is that Earth's axis has a large wobble and it takes 26,000 years to complete one of these wobbles. The result is that Earth's axis slowly changes the direction it is pointing in a large circular motion like a spinning top (and therefore continually changes its Pole Stars). With the completion of one wobble the axis has returned to its original position and once again points to its original pole stars. The Ecliptic remains relatively the same throughout this wobble but the direction that the axis is pointing will vary up to 49° (twice Earths maximum ecliptic 24.5°).

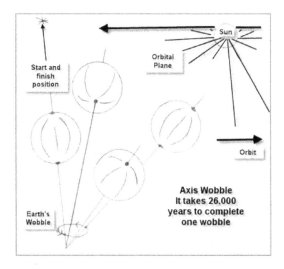

Figure 21 – Earth's Axis slowly changes direction due to "Wobble"

The most important consequence of (or result occurring from) this wobble is that the stars and constellations, if viewed from the same location over thousands of years, would be seen to cross the night sky progressively higher or lower. The angle above the horizon that any star is viewed at, as it crosses its meridian line, will change by up to 49° over thousands of years. This is the same as the maximum Earth axis change due to its wobble and is directly related. This star movement is called Precession. The completion of one wobble brings Earth's axis back to its original position whereby the stars are once again viewed at the original angle (above the horizon) that they had been at the start of the wobble. Also, Earth's Axis once again points at the original pole stars.

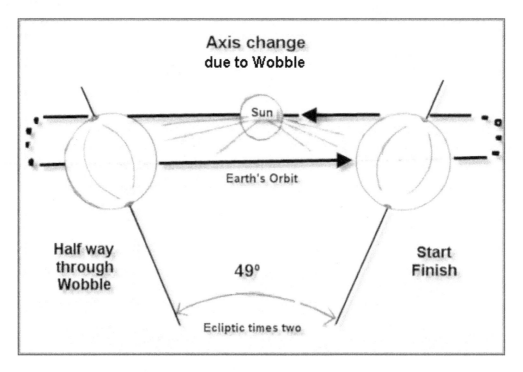

Figure 22 – "Wobble" causes the stars to change position (Precession)

Example: As the Earth's axis moves from its start point to half way through the circular wobble (49 degree change), a star will be seen getting progressively higher each century until it is ultimately 49 degrees higher. As Earth's axis returns to its start point of the wobble, the star now drops in the night sky to its original position. Each star will therefore only be at the same angle (at Meridian crossing) twice in 26,000 years, once as it travels up and once as it returns down.

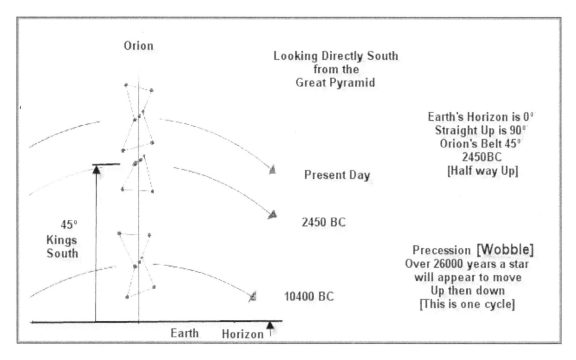

Figure 23 – Over time Orion crosses its meridian at different angles due to Earth's "Wobble"

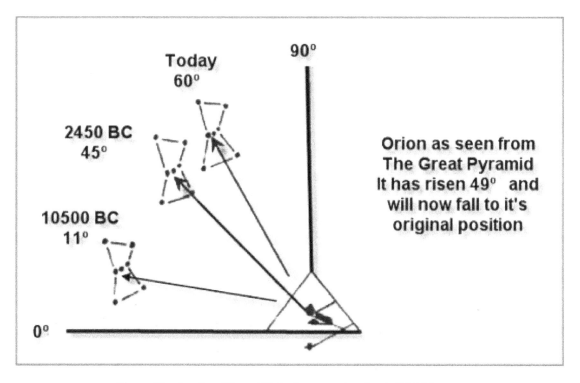

Figure 24 - Another View of Orion's precession due to Wobble

Upcoming in this book will be discussions concerning degrees of an angle and degrees of latitude. Although these are separate concepts, they are also essentially the same. These concepts are at the heart of The Great Pyramid.

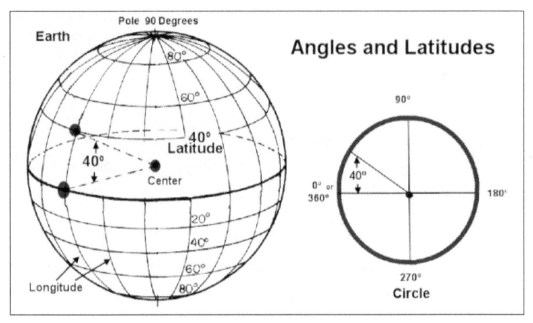

Figure 25 – The Ancients were mathematicians

One other very important discovery that Robert Bauval made was that the Giza pyramids also marked a much earlier time in history other than their date of construction. The pyramids were purposely positioned at Giza and in other locations so as to mirror the stars of Orion as they would have appeared in 10,500 BC. Precession not only raises and lowers constellations over time; it also continually rotates them back and forth. Bauval found that the angle in the sky that the three stars of Orion's Belt made in 10,500 BC (in relation to the horizon) best reflects the angle we see today with the three pyramids at Giza. It is no coincidence that the ancient Egyptians also marked this far distant date (8,000 years before the construction of the Giza pyramids), it was of extreme importance to them and was refereed to often in sacred texts as the "First Time". You will see that this earlier date is at the heart of this mystery and that The Great Pyramid confirms Bauval's 10,500 BC date in a second way -*it backs itself up*- so there is no misunderstanding! (Note-this second way best fits 10,400 BC and so this is the date used throughout the book)

Chapter 5

The Great Pyramid Speaks

The Great Stone Map

The high degree of architectural and engineering standards that were used to build The Great Pyramid were required in order to show a precise mathematical representation of the Earth's Northern Hemisphere. Imagine the Earth cut in half at the equator. The top half has been mathematically represented in a pyramid shape and scaled down. A similar upside-down pyramid can be imagined to represent the Southern Hemisphere. The outside base of The Great Pyramid represents Earth's equator and the top represents the North Pole.

To try and visualize this just imagine the four sides at the base of the pyramid slowly bulging out and the corners being pulled in until they formed a circle. At the same time the pyramid above was morphing the same way into a half sphere. This process has been mathematically built into the dimensions of The Great Pyramid. If you add the length of the four sides at the pyramid's base and use the resulting sum as the length of the circumference of our imagined half sphere, then the radius worked out equals the exact height of The Great Pyramid.

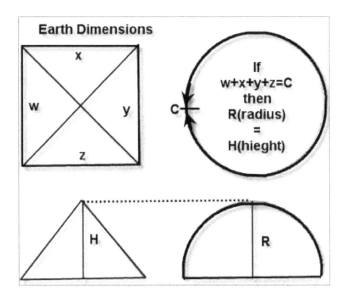

Figure 26 - "Squaring the Circle": Great Pyramid Models the Northern Hemisphere

This "squaring the circle" can be easily verified using the basic equation for a circle that we all learned in school C=2ΠR . Deep sockets were chiseled out of solid bedrock for each of the four corner stones. From these we have determined very closely the original base lengths and height of The Great Pyramid. The base between sockets was a square with sides of 764 feet and the height was 486 feet to the top of the missing capstone. So if C (circumference) is the length of the four sides then   C =4×764 =3056 feet and we know   Π (pi) = 3.14

$$C=2×Π×R$$

$$3056=2×3.14×R$$

therefore   R (radius)=486 feet (the height of the Pyramid)

Calculations such as these for the various aspects of the Pyramid's correspondence with the Northern Hemisphere are beyond coincidence. The Great Pyramid's dimensions reflect those of a half sphere and this half sphere represents Earth's Northern Hemisphere.

Was the ancient Egyptian's knowledge of the Earth so advanced that they knew there was a bulge at the equator? The centrifugal force of the Earth spinning around its axis causes this bulge and the Egyptians may have mathematically shown its existence in The Great Pyramid by constructing it with slightly concave sides. If this concavity was for structural purposes then why was it not used in the other massive pyramid at Giza?

Figure 27 – Concavities were first noticed from the air

The length of the concave sides at the pyramid's base is greater than it would be if all four sides were flat (straight). Did the Ancients know that the Earth's circumference at the equator is greater than at the poles? Now that would be precision, our advanced society has only recently discovered the existence of the equatorial bulge. How could such an early civilization have such advanced knowledge and why was representation of the Northern Hemisphere so important to them that they cut and moved over 15,000,000 tons of stone?

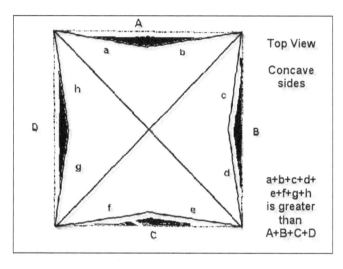

Figure 28 - Concave Sides of Great Pyramid (not to scale)

In order to continue this investigation, we must also bring in two other pyramids that play an important role to truly understand the advanced intelligence that we are trying to comprehend. These pyramids are named the Red Pyramid and the Bent Pyramid. They were constructed at the same time period as the three large pyramids at Giza and were similar in size although they are located at Dashur, some distance from the Giza pyramids. The existence of these two pyramids may be crucial facts that must be understood in order to understand the true extent of the complexity of this mystery.

Examining the construction of these two pyramids compared to the three at Giza it is very interesting that only three appear to incorporate the concave sides. Two of these pyramids are located at Giza and the third at Dashur. As stated above, if the concave sides of The Great Pyramid were symbolically used as representation of the planet Earth then what was the purpose of having concave sides on the small pyramid at Giza and the Red Pyramid at Dashur? One existing theory for these concave sides is that this was a type of construction used for stability of these pyramids yet the massive middle pyramid at Giza (as big as The

Great Pyramid) and the Bent Pyramid do not have concave sides. Knowing these pyramids were constructed within the same time period and using similar techniques, should we not expect that all these pyramids would have the same characteristics? Realizing the great intelligence at work here and understanding the extreme engineering difficulty involved to include such detail, we must try to understand the significance of these pyramids with concave sides.

Is the small offset pyramid at Giza, constructed with concave sides, a representation of both a star and the Moon? This pyramid's size in relation to the size of The Great Pyramid seems to be somewhat similar to the relation between the size of the Moon and the Earth. To think that this is even possible, we would have to know whether or not the ancients could have at least roughly determined the size of the Moon. Realizing their capacity to understand the solar system, it is reasonable to think that this information was gathered by comparing the size of the Moon's shadow cast upon the Earth (Solar Eclipse) and the size of the Earth's shadow cast upon the Moon (Lunar Eclipse). The question then: was it the Egyptians or an even more ancient civilization that obtained this measurement?

One of the most curious and intriguing pyramids is the Red Pyramid at Dashur, which was constructed with red stones and concave sides. Is it possible this pyramid marks the red planet we refer to as Mars? Knowing that the planet Mars has a red appearance, even with the naked eye, and appears to travel independently from the other stars, because it is a planet which orbits the sun, this planet would have been a very important "star" to the ancients that could have been used for the calculation of Time and Longitude.

Figure 29 – The Red Pyramid

When Christopher Columbus discovered America, he had no way of knowing how far he had sailed. While the calculation of Latitude is simply a function of how high the Pole Star is in the night sky, the calculation of Longitude is a function of Time. Not until the Eighteenth Century, with the invention of the Chronometer (large pocket watch), was this problem overcome. Before this many of the greatest minds of the time, including Galileo Galilei, Isaac Newton and Edmund Halley were actively trying to solve this problem and all thought the answer lay in the "clockwork" of the heavens. Did the ancients long ago solve the problem of Longitude calculation; was the movement of the planets the key? The change in distance of a planet from other stars can be measured for a 24-hour period (one Earth rotation) and then used as reference (time). Mysterious ancient devices have been found that some researchers believe were used for the accurate measurement of star positions.

There is yet another unsolved mystery, which pertains to the type of construction and location of the Bent Pyramid. This pyramid at Dashur, near the Red Pyramid, changes its angle of inclination as it rises above the desert floor. One thing that we understand more and more is that all of these pyramids are designed and built for a specific purpose and much more investigation will be needed to answer all the questions that remain. There is a story being told here of great importance and by ancients that were far more intelligent than is currently known.

Figure 30 – The Bent Pyramid

The Missing Capstone

In the ancient Egyptian culture the Benben stone was extremely sacred and was used to cap the most important monuments. According to mythology, the first Benben was a fallen meteor of conical shape. Indeed some large meteors found on Earth are of conical shape, this shape being caused by the extreme heat of entry into Earth's atmosphere.

It is said that the original Benben (meteor) was placed upon a pillar in the center of a courtyard that was surrounded on all four sides by large elaborate temple walls. This was the sacred Temple of the Phoenix located at ancient Heliopolis. Over time the Benben evolved into finely carved capstones that were placed with honor on the peaks of sacred pyramids and obelisks, often engraved with images of the Phoenix bird.

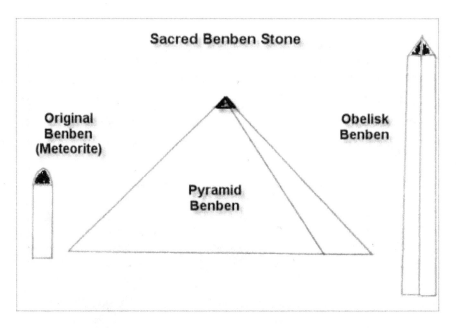

Figure 31 - Use of the Sacred Benben Stone

The myth of the Phoenix bird "rising from the ashes" is well known. It originated from the ancient Egyptians who considered this bird a symbol of rebirth and regeneration. Why did these Egyptians inscribe the Phoenix bird on the sacred Benben stone?

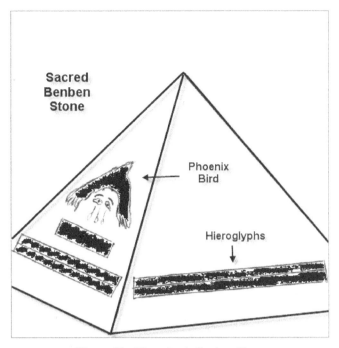

Figure 32 - Pheonix on Benben Stone

The missing capstone from The Great Pyramid was most assuredly a Benben stone. Knowing that this pyramid represents Earth's Northern Hemisphere, the capstone's position at the top would therefore mark the location of Earth's North Pole. Given all these clues and what we know about Egyptian symbolism, is there a message for us in the myth of the Phoenix? *Are we being told that the Earth was stuck at the North Pole in a cataclysmic meteoric event and that it was mankind that had to "rise from the ashes"?*

You may have some confusion at this point. You are now stepping "outside the box" of commonly accepted knowledge. Further chapters will corroborate this conclusion with facts that cannot be ignored. You will find that there is a great abundance of evidence that supports the message within The Great Pyramid.

The "King's Chamber"

Geographically the Great Pyramid (Giza) is located one third of the total distance from the equator to the North Pole. It lies at 30° N. Latitude with the North Pole being 90° N. Latitude.

A very important point that has been overlooked is the significance of the location of the Kings Chamber within the Great Pyramid. While constructing the Kings Chamber, a great amount of attention was given to the precise placement of the chamber so that it would be one-third the distance from the pyramid's base to the top of the capstone. *The King's Chamber was positioned in The Great Pyramid to represent Giza's position in the Northern Hemisphere at the time of the pyramid's construction (2450 BC).*

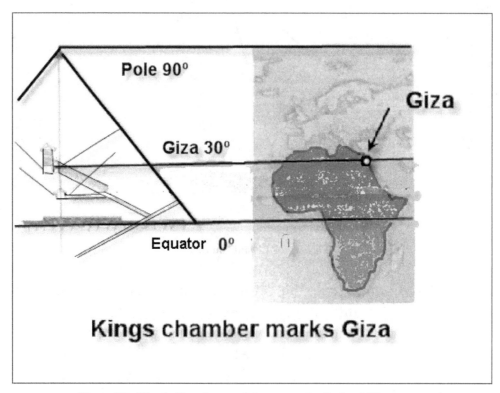

Figure 33 - King's Chamber models present Latitude of Giza

Directly above the Kings Chamber exists a massive structure built from the largest stones used in the entire construction of The Great Pyramid. Each one of the many stones used in this feature weighs approximately 70 tons. These stones were precisely fitted side by side in a beam type of construction for each of the structure's five levels.

Originally the only part of this structure exposed was the underside of the first level. This is the ceiling of the Kings Chamber. Explorers smashed their way through solid rock and discovered this beam structure with five related voids. Some authorities have speculated that this structure was positioned at that location to protect the Kings Chamber from the massive weight from above. The voids have since been referred to as the "Relieving Chambers".

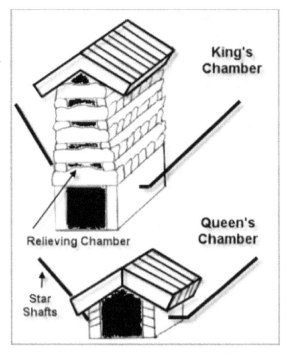

Figure 34 - Differing construction of King's and Queen's Chambers

Why is this structure only found above the Kings Chamber? Assuming this theory of weight relief is correct, was there no relief for the Queen, whose Chamber is located lower within the pyramid and therefore has the greater weight above. It seems clear, therefore, that the so-called Relieving Chambers were built to serve a purpose other than weight relief.

Was the precise construction of the Relieving Chambers used as representation of The Great Pyramid, is this structure self-referential? It appears to represent The Great Pyramid built within itself. Robert Bauval's theory on the Kings Chamber star shafts gave us a time in history (2,450BC). This new information gives us the position of The Great Pyramid (Giza) within the Northern Hemisphere at that time. *We now have our first "Time and Place".*

One thing that has commonly been suggested is that The Great Pyramid holds a secret of great importance. You will see that the marking of "time and place" is indeed of the utmost importance.

The "Queen's Chamber" — A New Interpretation

The Queens Chamber was carefully positioned in The Great Pyramid because it also represents Giza's position in the Northern Hemisphere, but Giza's position far back in the mists of time, her star shafts say 10,400BC (to be discussed shortly). The Queens Chamber is positioned a little more than halfway between the pyramid's base and the Kings Chamber. This positioning of the Queens Chamber within The Great Pyramid indicates a northern Latitude of 16°. The Queens Chamber is also positioned on the pyramids centerline while the Kings Chamber is purposely in an offset position.

The Great Pyramid is revealing that around 10,400BC there was a major impact event at the North Pole and as a result Giza's position on Earth was changed from its existing 16 degrees N. latitude to its current 30 degrees N. latitude. In this view it follows that The Queens Chamber indicates the geographical location of Giza at 16° N. Latitude before this catastrophic even. *We now have our second "Time and Place".*

As to the impacts of 10,400 BC, the consequences would have appeared in dramatic fashion on the Earth as it existed. To understand the changes in relation to Giza would be to understand the mammoth amount of energy that would have been released upon impacts. The Great Pyramid indicates that Giza underwent a violent repositioning of 14 degrees of latitude. There is only two ways that this could have happened, either the *entire landmass of Africa moved* or there was a *relocation of the Earth's poles.* Both of these scenarios could account for a previous positioning of Giza at 16 degrees N. Latitude.

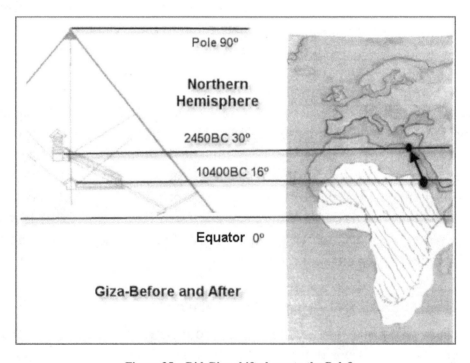

Figure 35 – Did Giza shift closer to the Pole?

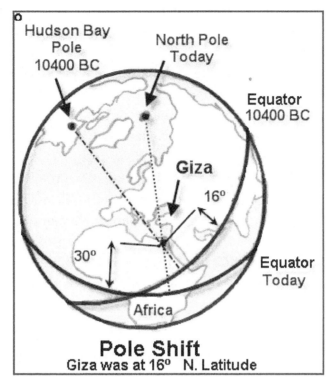

Figure 36 – Or did the Pole Shift closer to Giza?

As mentioned earlier in this book, even Albert Einstein found that the abundance of evidence suggested strongly to him that major Geo-Physical changes related to this time in history may have occurred. This date of 10,400BC, indicated by the Queens Chamber star shafts, is also within the time period associated with the end of the so-called "Ice Age" and a related mass extinction event.

As the structure above the Kings Chamber was used as a strong symbolization of The Great Pyramid, we must assume that the Queens Chamber and its different roof structure symbolize a smaller extremely important temple-like building that existed at Giza in 10,400BC. There is evidence that The Great Pyramid was constructed over top of much more ancient ruins, which may well have been an ancient star observatory. Was this earlier monument destroyed in the aftermath of an unimaginable impact event?

The Star Shafts – A New Theory

The Kings and Queens chambers each have two star shafts, one pointing North and one pointing South. We are proposing that both pairs of star shafts are pointing at the *meridian crossing* of the same two stars (Zeta Orionis and Alpha

Draconis) although at two vastly different times in history (2,450 BC and 10,400 BC) so as to show the change in star positioning that resulted from a catastrophic Pole Shift event (*"before and after"*).

The best way to understand the star shafts is to understand what we will call their "Gap Angle". For each chamber, the Gap Angle is the angle between the North and South star shafts. There are therefore two Gap Angles and they differ by only 1°: the Kings Chamber star shafts have a Gap Angle of 102.5° and the Queens Chamber star shafts have a Gap Angle of 101.5°. This small difference is very significant. Due to the Earth's wobble (Precession) the Gap Angle between the meridian crossing of our two referenced stars (Zeta Orionis and Alpha Draconis) continually changes. Any specific Gap Angle marks a specific time in history and this is how the Egyptians have shown us the timing of these catastrophic impacts (10,400 BC). The chart below tracks the Gap Angle between the meridian crossing of Zeta Orionis and Alpha Draconis back through time.

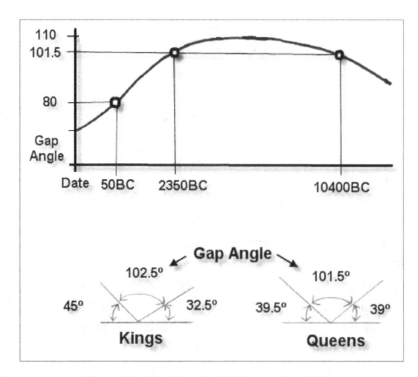

Figure 37 – The "Gap Angle" is a marker of time

Each pair of star shafts are marking the two points where a northern star (Alpha Draconis) and a southern star (Zeta Orionis) cross their Meridian Line. This

Meridian line is the line that would be centered directly above Giza's position on Earth and extend to two points on the horizon, one that is directly South and one that is directly North. These two stars (and all stars) daily cross this Meridian Line because the Earth is spinning. In the view presented here, both the King's and the Queen's star shafts mark these same two stars as seen from Giza but at two different times in history. These two different times in history each had a specific Gap Angle for our referenced stars and these are the two Gap Angles represented by the Kings and Queens chambers.

The distance between the two stars in the night sky does not change but the "slant" at which they cross the night sky does fluctuate during the 26,000-year cycle of the wobble. This is the reason the Gap Angle continually fluctuates and this is why the Gap Angle is an extremely accurate marker of "Time and Place". Investigation shows these two stars would almost never cross the Meridian Line at the same time each night and the amount of time between their crossings continually changes (depending on the slant). They might cross one hour apart at one time in history and two hours apart at another time in history.

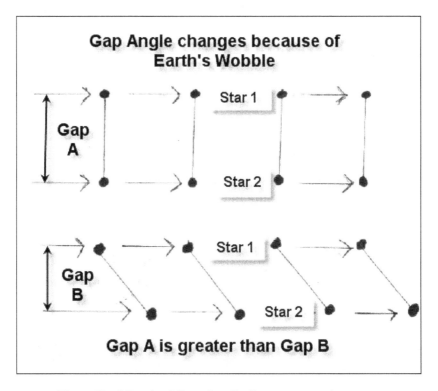

Figure 38 - "Gap Angle" continually fluctuates over time

If at any time in history you stood at Giza and pointed north and south at the two spots where these two stars crossed the Meridian line, the angle between your arms would be the "Gap Angle" related to that time in history. In other words, a specific Gap Angle indicates a very specific time in history.

Referring to the earlier graph we see that the Gap Angle in the Queens Chamber (101.5°) is only produced at two relevant times, 2,350BC and 10,400BC. The date 2,350BC is 100 years after the Great Pyramid's construction (2,450BC) and this approaching date may well have had great significance; however the date 10,400BC is the date the Queen's chamber star shafts are marking. In fact, you will soon see that the Egyptians have left little doubt that 10,400 BC was the date they were referring to.

Robert Bauval has shown that the layout of the three pyramids on the ground closely mirrors the layout of the three stars of Orion's Belt as they would have been seen in 10,400BC and it is known that this date had great significance in the ancient Egyptian culture. It was the sacred "First Time". So, if the position of the Queens Chamber within The Great Pyramid represents Giza's pre-impact position at 16° N. Latitude in 10,400 BC then it would make sense that the star shafts of the Queens Chamber would be marking Zeta Orionis and Alpha Draconis as viewed from 16° N. Latitude in 10,400BC. This is not difficult to determine, in 10,400 BC there was only one latitude on Earth from which Zeta Orionis and Alpha Draconis could have been viewed from at the angles indicated by the Queen's Chamber star shafts. Because of the curvature of the Earth, the same stars would appear at different angles from different latitudes. We should therefore first determine what latitude on Earth we would view Zeta Orionis and Alpha Draconis in 10,400BC so that they crossed directly north and south at the angles indicated by the star shafts in the Queens Chamber (was it 16 degrees N.?).

Using the Skyglobe program it quickly becomes clear that you would have only viewed Zeta Orionis at 39.5° and Alpha Draconis at 39° in 10,400BC from a point on 2° N. Latitude. Incredibly, *and of the utmost importance*, this latitude is also marked within The Great Pyramid. The ascending and descending passages intersect at a point that marks 2° N. Latitude given that this pyramid is a representation of the Northern Hemisphere. The chance this is a coincidence is very small and we hope to show you it approaches zero. The mathematical equation of what happened in 10,400 BC has been stated and answered in The Great Pyramid.

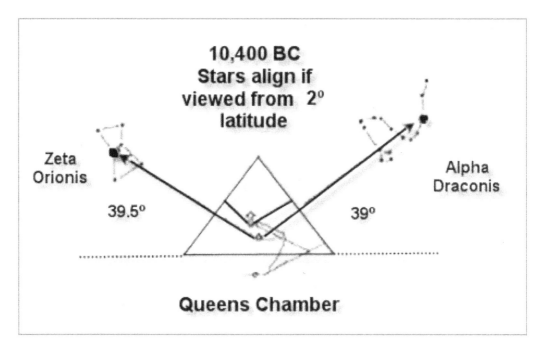

Figure 39 – Queens star Alignments for 10,400BC

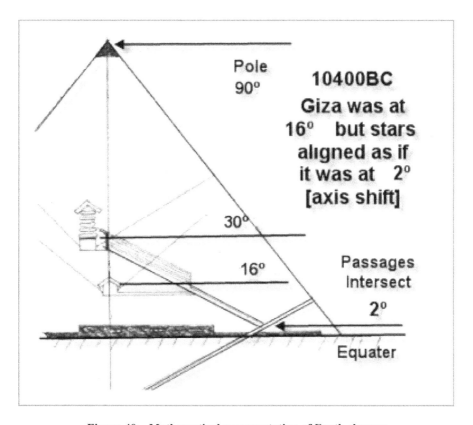

Figure 40 – Mathematical representation of Earth changes

The Great Pyramid is representing that in 10,400BC, just before impact, Giza was located at 16° N. Latitude and every night Zeta Orionis passed directly south at 39.5° and Alpha Draconis passed directly north at 39°. But Skyglobe shows us that you would only see these two stars at these angles from 2° N. Latitude in 10,400 BC. Given that both of these scenarios are correct, it appears The Great Pyramid is saying that at the same time Giza changed latitude by 14 degrees, the positioning of our referenced stars was changed by 28 degrees (the Queens chamber star shafts mark 2° latitude and the Kings mark 30° latitude- the difference being 28°), strongly indicating that some type of Earth axis change was involved. This information is very important in determining what exactly happened.

In constructing The Great Pyramid to represent the Northern Hemisphere it was fairly straightforward to show the latitudinal change of Giza (by the positioning of the two chambers). Representation of any axis change would be more difficult. In order for the Egyptians to show there was a change to Earth's Axis they constructed the star shafts to encode the amount of axis change (28 degrees).

Given the current star precession it can be determined (using star tracking software) that in 10,400BC, from a point on 30º N. Latitute in the Northern Hemisphere (Giza's present position), the southern star Zeta Orionis would have been viewed at 11.5º (Meridian crossing). But the related Queen's chamber star shaft indicates that this star was viewed at 39.5º (a 28 degree difference). One way to visualize what is being implied is to understand that if you physically rotated Giza directly south around the circular Earth by 28º, then both reference stars would lock into the Queen's Chamber star shafts.

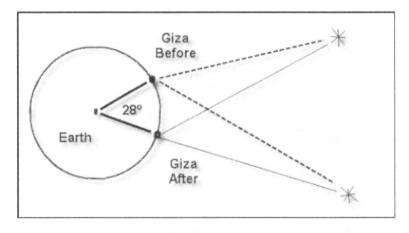

Figure 41 – Before and After

The Queen's chamber is marking the meridian crossing of two stars as seen from 2 degrees N. Latitude and the King's chamber is marking the meridian crossing of the same two stars as seen from 30 degrees N. Latitude. The difference (28 degrees) is representation of the total change in star positioning that resulted from some type of Earth Axis change.

It is hard to know if the Survivors fully understood what exactly happened, what we do know is that the survivors measured and recorded the changes that occurred and this information has been built into The Great Pyramid. At this point in the investigation, two basic assumptions can be made from the two basic pieces of information we have been shown.

1 – A 14-degree change in the position of Giza's latitude is recorded (by chamber positioning). Therefore: *Either Africa shifted closer to the North Pole or the North Pole shifted closer to Africa*.

2 - A 28-degree change in the positioning of two stars is recorded (by the star shafts). Therefore: Whether or not Africa shifted, some type of *Earth Axis change did occur* because even if Africa had shifted, it can only account for 14 degrees of the 28-degree star position change recorded by the star shafts leaving a further 14 degrees unaccounted for.

Although Giza was at 16 Degrees N. Latitude before impacts and 30 Degrees N. Latitude after impacts, the Egyptians marked 2º N. Latitude within The Great Pyramid for a very good reason: To show us *Earth Axis Change did occur*. Remember - the Queen's chamber star shafts were closed at both ends, completely entombed. Was this done to represent a lost time after which the stars would never be viewed the same again? Did this mark a new beginning? Was this the "First Time"?

Any Earth axis change would directly change the relative positioning of all stars because any axis change, be it internal (pole shift) or external (earth knocked over) would result in new Pole Stars and a change in Earth's rotation. Given their highly advanced knowledge, the survivors would have been able to calculate the amount of change to Earth's Axis by measuring the angle between the old Pole Star and the new Pole Star. They also could have determined Giza's new Latitude by measuring the angle that the new Pole Star was viewed at above the horizon (as they had previously done before impact to calculate Giza's previous Latitude).

There is only one way the Egyptians could have been aware of Giza's exact location in the Northern Hemisphere before this impact event 8,000 years earlier. Survivors of this catastrophic event must have passed down this information. At least one highly advanced civilization must have existed before 10,400 BC and they must have known Giza's previous latitude. This would have only been possible with great knowledge of the Earth's geometry and its movements as they relate to the stars; *they must have known that the Earth was round.* It was not that long ago that we believed the Earth was flat. This information confirms that at least one very highly advanced civilization existed many thousands of years before the Egyptians. Who were they?

Precession of the stars was different before and after the impacts of 10,400 BC. Forced to start from scratch, the new precession must have been carefully studied for the next 8,000 years and this was enough time to understand the new star movements. The ancient Egyptians then used this knowledge to explain the events of 10,400 BC in relation to the current precession.

The Ancients used the stars to locate their position on Earth in a similar manner to our use of the Global Positioning System of today. The surviving Ancients of this cataclysmic event continued with their precise charting of the stars in order to understand the new procession and star locations. The charts of old, which may have been based on many thousands of years of star observations, were now of little use. By the time of the Egyptian civilization the new procession of the stars was well understood due to fact that it had been recorded for 8,000 years previous.

The Ancients' great knowledge of Earth prior to impacts enabled the survivors to calculate the amount of change to Earth's Axis and Giza's relocation. This information (and a great deal more) must have been carefully preserved for thousands of years until it was lost some time after the building of The Great Pyramid. Fortunately, an early Egyptian civilization felt compelled to build a *Memorial* made of many millions of tons of stone and this monument has neither been lost nor destroyed. They knew well what it might be up against!

Suppose that we today were capable of lifting and moving this six million ton pyramid to a point of 2° Latitude and then used reverse procession to wind the star clock back to 10,400BC. Then amazingly, our reference stars would be at home as seen through the shafts of the Queen's chamber within this remarkable Great Pyramid of Giza.

Knowing that the reference stars Zeta Orionis and Alpha Draconis as viewed from Giza today do not line up with the star shafts of the Queen's chamber, we must turn the combination two times from this point backwards until these stars lock into place. We know that the Earth has a wobble and understand the resulting precession of the stars. From this we can obtain the first number of this combination. Precession causes movement of the stars over great amounts of time. This movement is not visible to us but we know that it causes the stars to slowly move up and down over thousands of years. Understanding precession, we can determine the position of our reference stars as seen from Giza at 10,400BC. We now slowly return the stars back to the click of the first number.

In order for us to make this combination click for a second and final time we must again backtrack the stars, this time by 28 degrees, to reverse the effects of axis change. We now see our reference stars moving ever closer to the exact position (angle) they had been seen at from Giza the night before impact (Zeta Orionis at 39.5° and Alpha Draconis at 39°). As star rotation reaches 28° we hear a very loud and final click. The stars are now locked directly in the star shafts of the Queen's chamber (of course the stars had moved little, it was always the Earth). You can follow these steps for our southern reference star, Zeta Orionis, in the following chart.

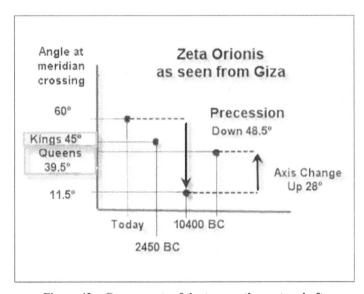

Figure 42 – Components of the two southern star shafts

The actual "Skyglobe" screenshots that follow show the Meridian crossings for Alpha Draconis and Zeta Orionis for an observer at 2° N. Latitude in 10,400BC.

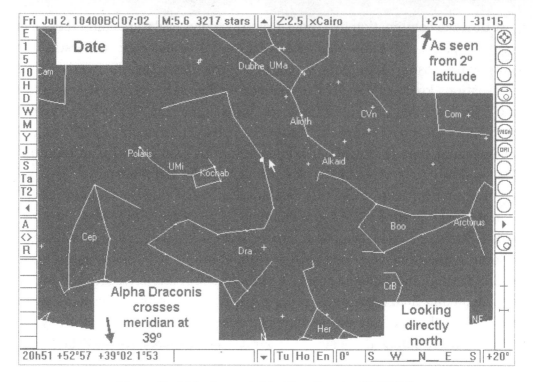

Figure 43 - Alpha Draconis Meridian Crossing 10,400 BC

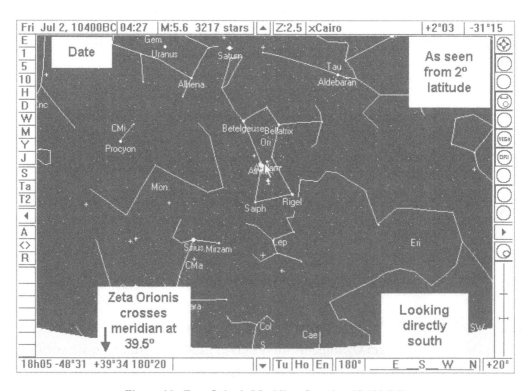

Figure 44 - Zeta Orionis Meridian Crossing 10,400 BC

The Causeways and Pyramid Positioning

One mystery that may well be related to the post-impact 28° change in star position relative to Giza is the curious orientation of the Causeways of the First and Second pyramids. These causeways are offset respectively 14° North and South of due East and so we see the two angles combine for a total of 28° in seeming correlation with previous findings.

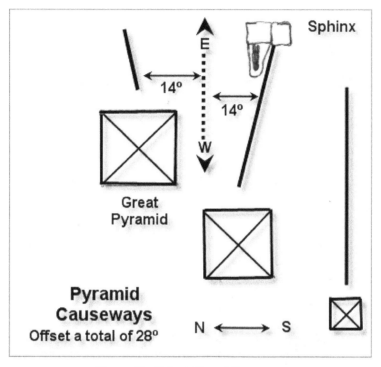

Figure 45 - Orientation of Causeways

As mentioned earlier the layout of the three Giza pyramids best matches the orientation of the Orion Belt stars as they would have been viewed at around 10,400 BC. Also at this much earlier date, the Sphinx would have been looking directly at its namesake constellation, Leo, as it rose above the horizon each night. After all of this, it is evident that there are too many points of fact for this to be thought of as a coincidence.

No explanation of The Great Pyramid is complete without an understanding of the largest and most magnificent chamber within The Great Pyramid. Perhaps the ramp, known as the Grand Gallery, with its oddly positioned slots and centered channel, was necessary in order to aid in the movement of the massive (70-ton) beams, which are positioned above the Kings Chamber. Envision a system of thick timbers and pulleys and levers. This may be another glimpse of the highly advanced engineering taught by the ancients.

Figure 46 – The Grand Gallery

If Giza and her pyramids were purposely constructed to record the catastrophic events of 10,400 BC, does this reflect on the meaning and purpose of pyramids constructed outside of Egypt?

In 1947 the New York Times ran an article entitled "U.S. Flier Reports Huge Chinese Pyramid". Colonel Maurice Sheahan had reported seeing, on a flight over China, a pyramid that would dwarf those of Egypt. Two days later a picture appeared in the paper. This story remained controversial for many years due to the fact that these areas of China were "forbidden zones" and independent verification was impossible.

Figure 47 – First hint of a Chinese connection 1947

Only recently has it come to light that China once enjoyed a rich history of pyramid construction and today a great many have been documented, some as ancient and as massive as those of Giza.

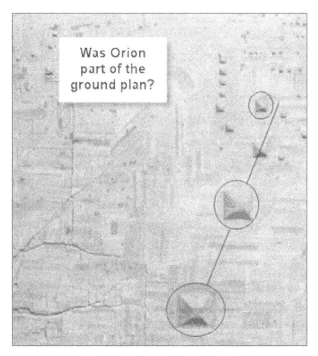

Figure 48 – Chinese Pyramids

Given the vast number of pyramids also located in Central and South America, the question again arises as to a possible connection. The largest such pyramid, the *Pyramid of the Sun* in Mexico, was built on near exactly the same footprint as the *Great Pyramid* (so close that estimates are just a few feet one way or the other). This can be no coincidence, these pyramids are connected!

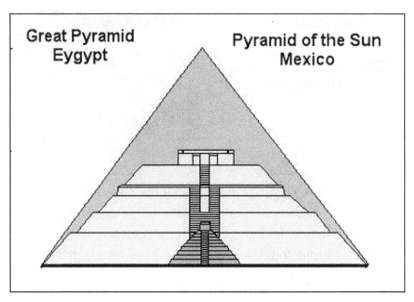

Figure 49 - Pyramid Sizes

This is strong evidence that ancient cultures, spread out over vast distances and across oceans, somehow shared important knowledge and culture. Are all these pyramids related to the cataclysmic impact event of 10,400 BC and its lost civilizations? Later in this book is more on the message within The Great Pyramid but for now please consider this; Robert Bauval found that the stars "Zeta Orionis" and "Alpha Draconis" were targeted by the star shafts of the Kings Chamber at the generally agreed upon time of the Great Pyramids construction, 2450 BC, and intriguingly he also found another date in the Giza pyramids layout, 10,400 BC. Now we find these same two important stars also align with the Queens Chamber star shafts at this much earlier date of 10,400 BC. *Understand this book is a result of a prediction that this Queens Chamber star alignment would occur if this pyramid was recording an "ice age" Pole Shift* or put another way *the chance this is coincidence is the same chance that you could pick two stars in the night sky and then predict where they were 10,400 BC!*

Chapter 6

The Cycle of Cosmic Catastrophes

Ice Age Impact

It is commonly thought that thick ice sheets covered much of North America until their rapid disappearance at the close of the last "Ice Age". What if these ice sheets only existed because the area of North America that was covered in ice used to lie within the Arctic Circle?

The Great Pyramid indicates that the Earth was struck near the North Pole in a meteoric event and the resulting impact was so cataclysmic that Giza changed latitudinal position. If this is correct, there should be at least one massive crater located in the northern Polar region although if the northern Pole was originally positioned on North America prior to impacts, then we must broaden our search area in order to find the supposed crater(s).

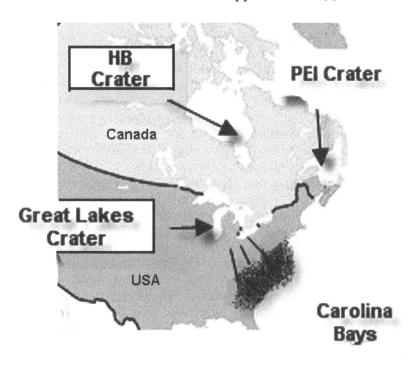

Figure 50 - The Craters

Some researchers have speculated that Lake Michigan and Lake Huron of the Great Lakes form a "horseshoe crater", a shape which is typical of a low angle impact. Also mentioned as being possible craters are three very large arc-like formations, one on the lower east coast of Hudson Bay, one surrounding Prince Edward Island on Canada's east coast and the last one in Finland. The problem is that none of these possible craters show rock deformation typical of previously studied craters and therefore no verdict has yet been possible.

Given the sheer size (these would be the four largest craters ever found on Earth), it is safe to assume that their signatures might well be unlike anything yet studied. These possible craters will be discussed shortly.

Other than craters, is there any scientific evidence that the northern hemisphere was struck about 12,400 years ago? Yes! There is very strong evidence that has been put forward recently by a respected scientist with the prestigious Berkeley Labs. Richard Firestone and his colleagues have put forth the theory that the Northern Hemisphere was struck by very large fragments of an exploding star (Supernova) and right in the time period we are discussing.

Here is a recap of his findings from a Berkeley Labs release:

A distant supernova that exploded 41,000 years ago may have led to the extinction of the mammoth, according to research that will be presented by nuclear scientist Richard Firestone of the U.S. Department of Energy's Lawrence Berkeley National Laboratory (Berkeley Lab). Firestone, who collaborated with Arizona geologist Allen West on this study, unveiled this theory at the 2nd International Conference "The World of Elephants" in Hot Springs, SD. Their theory joins the list of possible culprits responsible for the demise of mammoths, which last roamed North America roughly 13,000 years ago. Scientists have long eyed climate change, disease, or intensive hunting by humans as likely suspects.

Now, a supernova may join the lineup. Firestone and West believe that debris from a supernova explosion coalesced into low-density, comet-like objects that wreaked havoc on the solar system long ago. One such comet may have hit North America 13,000 years ago, unleashing a cataclysmic event that killed off the vast majority of mammoths and many other large North American mammals. They found evidence of this impact layer at several archaeological sites throughout North America where Clovis hunting artifacts and human-butchered mammoths have been unearthed. It has long been established that human activity ceased at these sites about 13,000 years ago, which is roughly the same time that mammoths disappeared.

They also found evidence of the supernova explosion's initial shockwave: 34,000-year-old mammoth tusks that are peppered with tiny impact craters apparently produced by iron-rich grains traveling at an estimated 10,000 kilometers per second. These grains may have been emitted from a supernova that exploded roughly 7,000 years earlier and about 250 light years from Earth.

"Our research indicates that a 10-kilometer-wide comet, which may have been composed from the remnants of a supernova explosion, could have hit North America 13,000 years ago," says Firestone. "This event was preceded by an intense blast of iron-rich grains that impacted the planet roughly 34,000 years ago."

"It's surprising that it works out so well", says Firestone.

Richard Firestone has recently released a new book based on his findings entitled "The Cycle of Cosmic Catastrophes". His theory is that our Solar System was struck by a massive amount of debris from this Super Nova and that perhaps all planets, not just the Earth, were impacted.

While investigating for ancient human settlements in North America, Richard made a very startling discovery. While excavating at widely separated sites he continued to find very high amounts of radioactive isotope Potassium 40 in human artifacts and soil at ground layers closely associated with 10,400 BC, (he relates 10,900 BC). The interesting fact about Potassium 40 is that it is not naturally found on Earth at these levels but it is created in large quantities by exploding stars (supernovas). This Potassium 40 appears to have rained down on the surface of North America in close proximity to 10,400BC. Another important fact is the lack of human artifacts found above this layer of Potassium 40, which is strong evidence that human activity ceased after this event for a very long time. After following up on these and other clues Richard Firestone came to the strong belief that the Northern Hemisphere had in fact been blasted by massive fragments from a supernova.

You will find in this book that many points of fact will be discussed that taken on an individual basis are powerful evidence that something catastrophic happened around 10,400 BC, although there has always been difficulty in pinning down exactly what. The Great Pyramid now allows us to understand this event as never before and this information should be considered and not ignored.

Further confirmation comes from ice core samples taken from Greenland's ice sheets. Each year a layer of ice was laid down so that reading ice cores is like reading tree rings, each layer representing one year. Independent researchers have determined that the element Iridium settled on these ice sheets around 10,400BC. Iridium is known to be related to extraterrestrial impacts and Iridium layers are used to date meteor impact events. This was the major evidence put forward as proof that the Dinosaur extinction was a result of an impact event. In many locations around the world where 65 million-year-old sediment levels are exposed the element Iridium is found at very high levels. No Dinosaur bones are found above this layer. Knowing that the ice sheets of Greenland are the only remaining northern ice from the period known as the "Ice Age", this is therefore one of the places that was able to preserve the iridium evidence and record a very accurate time for this catastrophic event. It is the same time that the great Egyptians have marked in The Great Pyramid as the date of a major impact event.

The Hudson Bay Crater

A particularly interesting formation can be found at the lower eastern shore of Hudson Bay. This coastline represents one half of a near perfect circle and is completely filled with lava. A few scientists have referred to this as a possible impact crater due to the knowledge that most meteor impacts will form circular craters and such formations are extremely rare in nature.

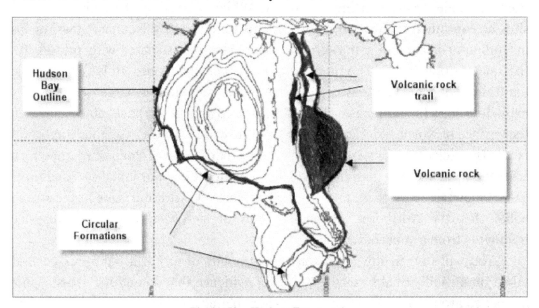

Figure 51 – Hudson Bay crater

A perplexing mystery remains as to the location of the missing half of this crater and its lava. Also, why do channels of lava lead away in a northerly direction from this location? Could this trail of lava indicate major land movement within North America? Does this trail lead back to the other half of this crater, now crushed and hidden?

There are many anomalies in and around Hudson Bay that indicate this is a "bullet hole" crater and that something massive pierced North America at Hudson Bay, cracking the continent and allowing major land movement to shift the two halves of the crater apart. Lava must have quickly filled the "bullet hole" and then left a lava trail as one half of the crater was shifted in a northern direction. The remaining half was not damaged and is still perfectly circular (as only craters have been shown in nature) and is in pristine condition. There is no way this perfect crater shape could of survived under miles of moving ice for millions of years of the so called Ice Age as science states, it would be scarred and disfigured. No this crater is post ice; this crater is why the ice disappeared!

In order to understand the amount of geographical changes that occurred, given this is a meteor crater; we must understand that the impact zone under discussion caused reconstruction of the Earth's surface in the form of a crater 500 km wide. Try to imagine standing at the edge of a crater this size. This will make it easier for us to understand the tremendous amount of debris that was blasted out upon impact in many different sizes and shapes and travelling up to 1,000 miles or more. In order to understand the force of this impact correctly, if we were to assemble all of the world's nuclear weapons in one location and detonate them, this would not come close to the amount of force released by this major impact. We must further understand that the meteor we are discussing may have been almost 500 kms wide (a bullet is not much smaller than its hole) and travelling up to 30 times the speed of sound. Truly understanding the devastation caused by such a large meteor makes it more conceivable to understand the drastic changes that The Great Pyramid attests too.

For years scientists have been debating the placement of Erratic Boulders. These are boulders found lying upon the ground in areas away from where they originated. They are found by the millions in North America and Europe and vary in size from the smallest to some as large as apartment units.

Figure 52 – Six foot tall Erratic Boulder

Science has tried to show these erratics were moved by glaciers during the last "Ice Age" although could they not have been propelled great distances by meteor impacts? A very strong fact that leads in the direction of rock propulsion is the recent discovery of unique black erratic rocks called Omars in North Dakota, USA. The only known source of this rock type is about 1000 miles away at the Omarolluk outcrop on the Belcher Islands, which is located within the lava field of Hudson Bay.

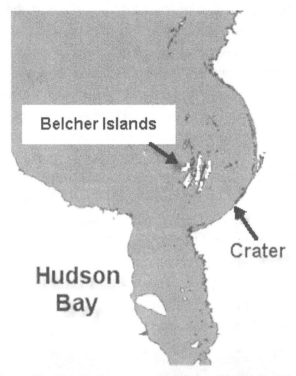

Figure 53 – Only known source for "Omars" is Belcher Islands

It is an accepted fact that throughout history glaciers have only been shown to move in a downhill direction (or short distances over level ground). Due to geographical conditions it would be virtually impossible for these Omars to have been carried "uphill" by glaciers for such a long distance. It is quite possible that these Omars and millions of other erratic boulders were repositioned by large meteor impacts.

Is this Hudson Bay crater the impact site referred to by the Ben Ben stone? Remember, The Great Pyramid states that the Earth suffered a great impact near the North Pole. Was Hudson Bay the location of the North Pole before the impacts of 10,400 BC? Charles Hapgood (*Path of the Pole*) believed this was indeed the location of the North Pole before about 12,000 years ago. If so, there would be no doubt as to the existence of large ice fields encompassing the Hudson Bay region and under these conditions this catastrophic impact would have occurred on vast polar ice sheets. Understanding the amount of heat and energy released by this impact, it is reasonable to assume this would have caused massive movement and rapid melting of large portions of these ice fields, resulting in mammoth floods.

Science is in agreement that mega-floods did occur around the end of the so-called Ice Ages although the mechanisms and timing methods used to explain these floods are suspicious. All of science's evidence put forward has never once considered impact until recently and the timing of such mega-floods has relied heavily on dating methods such as Carbon 14 dating. Richard Firestone has shown *these impacts would have greatly altered atmospheric conditions and such dating methods as Carbon 14 could well produce dates that are many thousands of years in error.* Putting aside the timing of each flood; it is more important you understand the severity of these events. Here are four examples showing that massive amounts of floodwaters did originate from northern North America and surged in all directions.

South

So much fresh water was rapidly released down the Mississippi flood plain that the ocean waters contained within the Gulf of Mexico went through a profound chemical change in a short period of time. Perhaps this helps explain the strong evidence that the Grand Canyon was carved quickly and recently, discarding sciences farfetched explanation of how a small river flows up to a high plateau wall and then magically carves a canyon through it.

East

We know one flood down the Hudson River Valley into the Atlantic Ocean contained 15 times the volume of all the rivers in existence today. This flood gouged a canyon in the continental shelf and left truck sized boulders strewn about the ocean floor (imagine the energy needed). It is thought this one flood alone raised ocean levels.

West

The energy released in the Missoula floods moving towards the Pacific Ocean through today's Columbia River Valley carved slightly smaller versions of the Grand Canyon and devoured vast quantities of land in miles-wide channels referred to today as the Scablands.

North

The northern floodwaters tore gigantic pieces of ice field apart and carried them down the Greenland Strait towards the equator where the evidence of mega ice-burgs grinding to a halt on the Ocean Floor can still be seen today all the way along this corridor and down the east coast of the USA.

Whether directly or indirectly, these floods were a result of the impacts of 10,400 BC. To further help us understand the magnitude of these impacts and realize that the origins of many ancient myths may have been based on actual events, we can look at the myths put forward by the American Indians which have been passed down through countless generations. As well as flood, a common element to many American Indian myths was the association of incredible heat due to this cataclysmic event. This amazing tale is from the Brule Indians, members of the Lakota Nation.

From "The Cycle of Cosmic Catastrophes"

Battle of the Giant Animals
Retold from Erdoes and Ortiz, 1984

In the world before this one, the People and the animals turned to evil and forgot their connection to the Creator. Resolving to destroy the world and start over, the Creator warned a few good People to flee to the highest mountaintops. When they were safe, the Creator sang the Song of Destruction and sent down fierce Thunderbirds to wage a great battle against the other humans and the giant animals.

They fought for a long time because the evil humans and the animals had become very powerful, and neither side could gain an advantage. Finally, at the height of the battle, the Thunderbirds suddenly threw down their most powerful thunderbolts all at once. The fiery blast shook the entire world, toppling mountain ranges and setting forests and prairies ablaze. The flames leapt up to the sky in all directions, sparing only the few People on the highest peaks. It was so hot that the world's lakes boiled and dried up before their eyes. Even the rocks glowed red-hot, and the giant animals and evil people burned up where they stood.

After the Earth finished baking, the Creator began to make a New World, and as the Creator chanted the Song of Creation, it began to rain. The Creator sang louder and it rained harder until the rivers overflowed their banks and surged across the baked landscape. Finally, the Creator stamped the Earth, and with a great quake the Earth split open, sending great torrents of water surging across the entire world, until only a few mountain peaks stood above the flood, sheltering the few People who had survived.

After the waters cleansed the Earth and subsided, the Creator sent the surviving People out to repopulate the new world, our world today, warning them not to fall into evil, or the Creator would destroy the world again. As the People went out over the land, they found the bleached bones of the giant animals buried in rock and mud all over the world. People still find them today in the Dakota Badlands.

While excavating for answers, Richard Firestone discovered that throughout North America and much of Europe there exists a "Black Mat Layer" at soil levels that date back to the era of 10,400 BC. One of the properties common within this "Black Mat" was a charcoal like substance indicating that all types of matter were incinerated by some continent wide inferno. These impacts we are discussing would have caused continent wide temperatures capable of reducing all existing matter to this charcoal like substance.

Besides intense heat, ancient myths often refer to incomprehensible hurricane winds, the appearance of the "sky falling" and torrential rain consisting of rocks. All of these things are consistent with catastrophic impact by meteor. Given these facts, surely the myths surrounding Atlantis may have some credibility.

The Great Lakes Crater

The formation of the Great Lakes in North America, as determined by science, occurred around the close of the last "Ice Age". The one thing science has not explained convincingly is just how these massive lakes were formed. The established theory is that advancing ice sheets gouged these lakes out. Is it not possible that a massive meteor is responsible?

It is documented that low angle meteor impacts carve out craters similar to that of a horseshoe. This is caused by the tremendous sideways release of force upon the Earth's structure and the direction of travel is shown. Together Lake

Michigan and Lake Huron of the Great Lakes are a prime representation of an immense horseshoe crater. The meteor would have been travelling in a southern direction like an airplane preparing to land.

Figure 54 - The Great Lakes

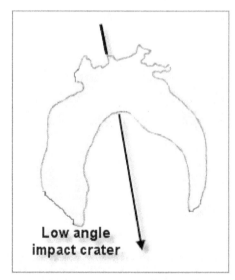

Figure 55 - Actual Horseshoe Crater

Also of importance are the curious circular rock patterns that spread out from a central point between these two lakes. This entire area (The Michigan Basin) has been compressed in a bowl-like formation and perhaps is indicating that a sudden and catastrophic compaction of the Earth's crust caused deep layers to buckle upward on the edges. Has science yet to recognize the signature markings of a massive meteor air burst? Are the Michigan Basin and other Basin formations representative of large meteor detonation above land?

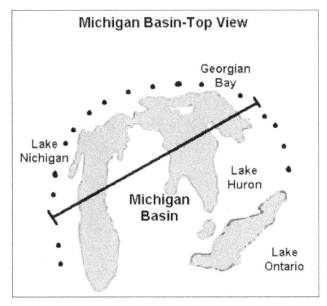

Figure 56 – Circular "Bowl" formation

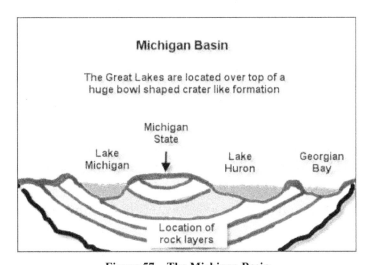

Figure 57 – The Michigan Basin

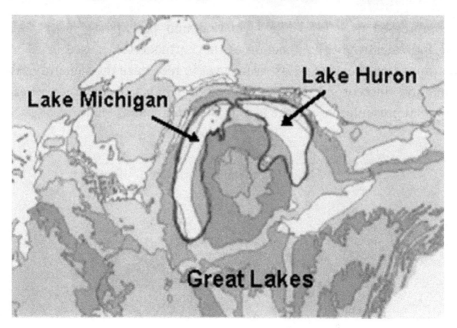

Figure 58 – The Great Lakes Crater

Also located within the Great Lakes region we have Lake Superior, the largest of the Great Lakes. The formation of Lake Superior, according to science, took place near the close of the so-called Ice Ages along with all of the Great Lakes. Is this an indication that it too was formed by meteor impact? What is very interesting is that Lake Superior also shows a somewhat basic horseshoe shape and also shows the same direction of impact as the Lake Huron/Lake Michigan Crater (as would be expected).

The Carolina Bays

A very important piece of information that adds credence to this theory of meteor impacts at the Great Lakes and Hudson Bay is the existence of the Carolina Bays. Spreading down the East Coast of the United States from New Jersey to Florida are small bay formations numbering maybe in the millions. These crater-like bays are all elliptical in shape and therefore show direction of impact. Elliptical craters have only been shown to be formed when water (liquid) impacts the ground at an angle. This information strongly helps to validate that major meteor impacts did occur on the miles thick ice fields of North America. Such impacts would have propelled massive blocks of ice into the atmosphere, which in turn could have been transformed to a liquid state due to the incredible heat of impact.

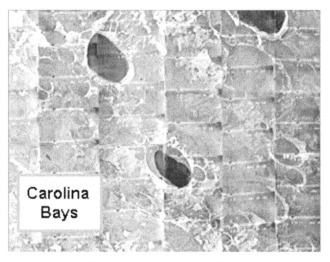

Figure 59 – Arial view of Bays

Very important is the positioning of these bays. All point back to the Great Lakes or Hudson Bay. The axis of these bays are divided into two distinct patterns, one group is pointed back to a central location at the Great Lakes and the other group points more northerly to a central location near Hudson Bay. Common to both groups is their consistent orientation; in Florida they point more northerly while as you move up the coast to New Jersey they increasingly point more in a western direction. The alignment to the Great Lakes and Hudson Bay remains constant.

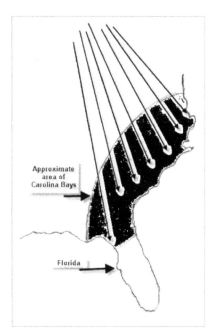

Figure 60 – Bays point back to Great Lakes or Hudson Bay

This is strong evidence leading to the conclusion that the Carolina Bays were formed as a result of large fragments of ice and water being propelled outward from two separate impact zones, both of which were originally covered in thick polar ice sheets. This also confirms that the "Ice Age" ice sheets of North America did exist and as we know the Greenland ice sheets are still present, so let's turn our attention to Europe and its proposed ice sheets of the "Ice Ages".

The Finland Crater

An important point to consider is the discovery of a possible massive crater located in Finland. The explanation given by science for this giant arc formation is that it was formed by glacial movement during the Ice Ages. Another explanation has been offered in *"The Cycle of Cosmic Catastrophes"* which states this formation was likely caused by meteor impact. Also entering this debate is the question "did ice fields really even exist in Europe?" D.S. Allen and J.B. Delair (*Cataclysm*) have shown that much of the evidence used by science to conclude thick ice cover in Europe can also be attributable to mega-floods and enormous thermal nuclear blasts (impacts). Is it possible that Europe was covered in vegetation, not ice, during the so-called "Ice Ages"? The common thread in this debate is the time period this giant arc was formed, which is 12,000 to 14,000 years ago. This time frame is in accordance with the date given by The Great Pyramid for a major impact event.

A vital clue is that this possible crater is closely centered at the point of maximum isostatic rebound in Northern Europe. Science's explanation for this rebound is that the land was greatly depressed by weight of massive ice fields during the period known as the Ice Ages. The theory goes that upon these ice fields melting, the compressed land began to rebound upwards. We know this land is still rebounding today. Given the isostatic rebound is centered on this possible crater, is it not possible meteor impact or air burst and not weight of ice fields cause the compression of this land? Did detonation force this landmass down into the Earth from which it is still rebounding today? While considering this point remember we do not definitively know whether ice fields even existed here.

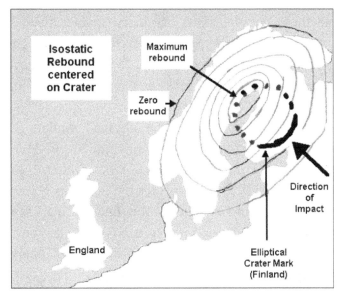

Figure 61 – Isostatic rebound centered at Finland Crater

While investigating the other main area of isostatic rebound, which is situated in North America, we find this rebound is also centered on a huge arc formation, the previously discussed Hudson Bay Crater. This appears to be an incredible coincidence, or is it?

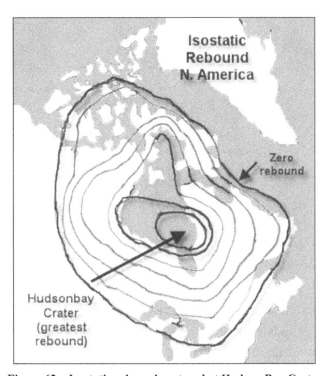

Figure 62 – Isostatic rebound centered at Hudson Bay Crater

Is it not possible these compressions happened instantaneously upon meteor impact and not over thousands of years as offered by science? The core scientific evidence in no way discounts this possibility. If these compression points were created instantly by impacts, the entire planet would have been dramatically affected.

In order to try and comprehend the physical damage caused by large impacts we can refer to the planet Mars. On one side of Mars we can see a massive impact crater well over 1000 miles wide (Hellas Impact Basin) and directly across on the other side of the planet the land has been bulged 10 miles high above surface level. Stretching across this bulge is a massive crack (Valles Marineris), which is close in appearance to the Grand Canyon but many times larger. These are all indicators of the massive amount of energy released and transferred through a planet by such events.

Clear and Present Danger

Another intriguing circular land formation is located on the eastern coast of Canada and surrounds Prince Edward Island.

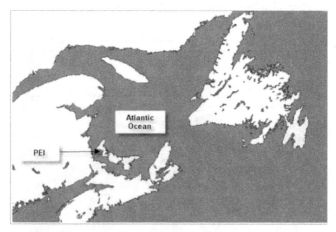

Figure 63 – P.E.I. crater

It has been speculated that this location could be yet another impact crater represented by the circular coastline. Another important fact for this equation is that seismic readings taken in 1950 indicated the possible existence of a deep cavity within this circle that is representative of meteor impact.

Because the evidence left behind at these possible impact sites is not typical of previously studied impact zones, scientists have viewed them with suspicion and little real investigation has occurred. Given that the sheer size is exponentially greater than anything yet studied we might expect the Earth to react differently under such extreme force. Let's now expand this investigation.

Another major impact event, one that appears to be even more recent than the ones being discussed now, occurred in the Indian Ocean only about 5,000 years ago. If the Holocene Impact Working Group is right about the time frame, this impact took place just 500 hundred years or so before the construction of The Great Pyramid. This impact event would have made it clear to the guardians of ancient knowledge that the events of 10,400 BC, now over 7,000 years in the past, were as the Ancients recorded.

To understand the devastation of this relatively recent event is to understand that a large meteor penetrated through deep ocean waters and created a large crater (Burckle Crater) on the ocean floor. *This impact had enough force to relocate "the ocean floor" up and onto the Island of Madagascar.* Four enormous ocean floor sediment deposits, called chevrons, have been found on Madagascar and all point back to this crater under the Indian Ocean. Each is the size of a large city and over 600 feet high. The fact that metals, which are typically formed by meteor impact, are found within these chevrons, we should have no doubt *this is an impact zone.*

Figure 64 – Ocean floor sediment chevrons on Madagascar

This incredible impact must have created giant tsunamis (minimum 600 feet) throughout the Indian Ocean and devastated shorelines worldwide. Climate records show that a sudden worldwide climate change occurred (dramatic global cooling). To put this in perspective, if this event happened unexpectedly today, all humanity could be at risk. Massive amounts of chemicals, nuclear waste and other pollutants would enter our oceans and atmosphere and the production of food, fuel and almost everything else would be greatly interrupted.

An important point that should be discussed is the timing of this enormous impact that occurred in the Indian Ocean and the apparent beginnings of the Egyptian, Sumerian and Indus Valley civilizations, which are considered by many the first truly advanced civilizations. All three of these civilizations appear to have started well past infancy stage at around 5000 years ago (3000 BC), the same time frame as this impact we are discussing (also the mysterious "A" Group civilization seems to vanish right around this same time period). Each of these civilizations were in close proximity to the Indian Ocean and the documentation shows that from the beginning these civilizations were complete with their own language and writing skills, religious beliefs, advanced building techniques, and sophisticated astronomical knowledge. This only makes sense if there was in previous existence highly advanced civilizations that were forced to relocate.

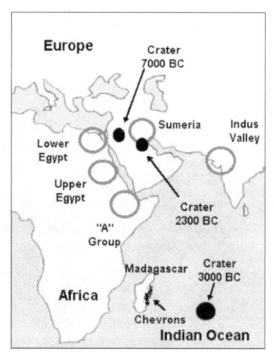

Figure 65 – Ancient Middle East

There is little concrete evidence supporting the existence of a former advanced civilization, yet there is much circumstantial evidence that points to their existence. Is it possible the physical evidence of a previous advanced society was destroyed and buried by this impact in the Indian Ocean and may never be discovered? We must understand that if a previous civilization was situated on the Indian Ocean prior to impact, it would have suffered total devastation (For example, if it was situated on the eastern coast of Madagascar then it may currently be buried under 600 feet of ocean floor sediments!). Assuming there were survivors, their relocation would have been necessary. Did these survivors join or conquer the inhabitants of the Nile Valley and other regions, and bring with them their advanced knowledge? Given The Great Pyramid, it is evident that advanced knowledge has been passed down from the very earliest of times.

Another piece of interesting evidence that may indicate these impact events and air blasts (atmospheric detonations) occur more frequently then we have been made aware of was found in the Pharaoh's tomb of Tutankhamun (King Tut). Mounted in the center of a priceless jeweled chest plate was a large glass stone. This stone has been studied and was shown to have been formed when desert sand was exposed to a great meteor air blast. This blast created such enormous heat it turned the sands of the Sahara desert to glass.

Figure 66 – King Tut's Scarab

Oceans cover more than two thirds of the Earth and therefore most impact events could be expected to have occurred over water. Direct evidence of such an impact would be difficult to find so we have no way of knowing how often these events have been happening. We should be concerned though as this type of impact may well be more devastating than a land impact. Even a relatively small meteor could unleash massive tsunamis that would endanger heavily populated coastlines.

A more recent and well-documented impact event occurred on the planet Jupiter. The comet Shoemaker-Levy 9 had broken into many large pieces, and these were photographed impacting Jupiter in rapid succession. Each impact caused a visible black spot larger than the size of the Earth. Can we really allow ourselves to believe that we were just lucky enough to get the chance to witness a once in a million years event!

In 1908, out of Siberia, came reports of a massive explosion. Only after preparing for many years was a team of investigators assembled to travel to this very isolated location. After travelling through heavily wooded areas for long distances the location of the blast was easily recognizable by its completely devastated and flattened condition. An area encompassing over 2,000 square miles showed evidence of complete destruction. Many years of debate led to the conclusion that this devastation was caused by a meteor about 300 feet long, which had exploded in Earth's atmosphere before it impacted the Earth's surface. To properly understand the devastation caused by this relatively small meteor would be to compare it to the destruction of Hiroshima by only one nuclear bomb. The explosion that took place over Siberia was equivalent to 2,000 nuclear bombs as used at Hiroshima.

Figure 67 – Tunguska impact – Siberia 1908

Understanding the damage caused by this meteor about the size of a football field, it would be impossible for us to comprehend the cataclysmic effect that a meteor many times larger than Mount Everest would have. We don't have to guess; the Egyptians have historically recorded the geo-repositioning of the continents that resulted by just such an impact within The Great Pyramid of Giza. Understanding that 12,400 years ago the Earth was devastated by meteors and only 100 years ago the Siberian meteor event took place, we must ask ourselves why science has such little understanding or interest in these facts.

Shortly after the construction of The Great Pyramid there is a large historical void in Egyptian history (the first Intermediate Period). Simultaneously, civilizations throughout the Middle East, Europe, the Indus Valley and as far away as China appear to have collapsed. There is substantial evidence to confirm that these areas suffered drastic climate changes and large amounts of devastation around 2,300BC. Only after maybe hundreds of years did civilizations re-emerge. Could this be yet another example of a recent meteor impact that we are yet to understand? Recent evidence does indicate this possibility. A highly recommended website that documents this evidence is SIS (Society for Interdisciplinary Studies – www.sis-group.org.uk). The SIS was formed in the belief that varied disciplines (such as biology, geology, astronomy and many others) should work together in order to better understand what role cosmic catastrophes may have played in Mankind's history.

There are still further indications that these impact events have been far more common than has previously been acknowledged. The identification of a large crater (the Jobal Waqf es Swwan crater) in Jordan (Middle East) has led to an estimated impact date of around 7,000 BC. Given the crater's size, it has been reasoned that the energy released by this impact would have been equivalent to 5,000 nuclear bombs as used at Hiroshima. Atmospheric temperatures over a wide area would have been in excess of 1,000 degrees C and millions of tons of debris would have been blasted into the sky, blocking sunlight for quite some time and greatly affecting climate. Climate change seems to be a reoccurring theme in association with impact events and civilization decline. Evidence of significant climate change is associated with other well known "times" such as the Second Intermediate Period in Egypt (1,800 BC) and the Dark Ages (540 AD). A surviving letter from a Roman Praetorian from around 540 AD describes the strange events:

Praetorian letter

"All of us are observing as it were, a blue colored sun; we marvel at the bodies which cast no shadow.... So we have had a winter without storms, spring without mildness, summer without heat.... The seasons have changed by failing to change; and what used to be achieved by mingled rains cannot be gained by dryness alone".

The great Roman civilization collapsed shortly after this letter was sent, in fact around the same time, but on the other side of the planet, the great civilization at Teotihuacan in Mexico also collapsed. The evidence suggests that for up to two years the sun was blocked worldwide by a haze of debris, which led to severe crop failure and the onset of disease such as the Bubonic Plague. The Dark Ages really did start off dark. Early investigators reckoned there must have been a colossal volcanic eruption yet no geological evidence for this has ever been found. More recent investigations have concentrated on the very real possibility of a comet or meteor impact.

Here is an excerpt from a recent New York Times article:

"Scientists in the Holocene Impact Working Group say the evidence for impacts during the last 10,000 years is strong enough to overturn current estimates of how often the Earth suffers a violent impact on the order of a 10-megaton explosion. Instead of once in 500,000 to one million years, as astronomers now calculate, catastrophic impacts could happen every few thousand years. This year the group started using Google Earth to search around the globe for chevrons, which they interpret as evidence of past giant tsunamis. Scores of such sites have turned up in Australia, Europe, Africa and the United States."

Perhaps this information will take Mankind from its complacency of thinking the last major impact was the time of the Dinosaur Extinction (65 million years ago). We must now accept that this is a much more common occurrence than we have been made aware of. Should more time not be spent to investigate the massive implications and to develop strategies to protect ourselves? Are we cutting our lawns with lions walking around our yard?

Chapter 7

10,400 BC

Pangaea

A scientific fact that has been established and accepted is the movement of continents. This fact has been established with the knowledge that there once existed a super continent, which we now call Pangaea. It is thought that this super continent broke apart into the continents we now know beginning over 200 million years ago. This theory began to take shape when it was noticed that some continents, when put together, appeared to mesh and when this was done, the ancient fossil tracks were found to align as well. Also supporting this theory is the existence of ancient mountain ranges that appear to once have been united but now shows signs of major separation and are located on different continents. Since the separation of Pangaea these continents have shifted to their present positions.

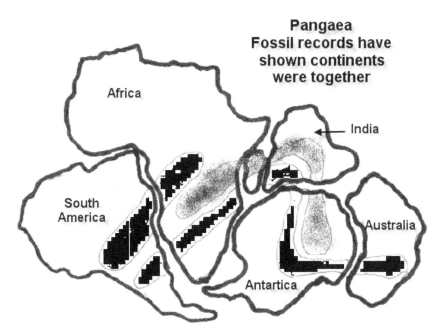

Figure 68 – Continents were once joined

Science states that there is slow constant movement of the continents. It is thought a very slow shifting of Earth's crustal plates causes this movement, which has been measured using modern devices. Science tells us this is how the continents have come to their present positions. In simplified terms the theory goes like this: the Earth's crust is made up of six large plates and a number of smaller ones, and each continent rests upon one of these plates or is cradled within it and therefore movement of a plate results in movement of the continent. In some places the plates are separating and the resulting gap that opens between them is continually filled with new lava. This new lava is forced upward and has created an incredible underwater mountain range, called the Oceanic Ridge, which winds its way around the globe like stitches on a baseball. In other areas the plates are colliding and one plate is driven down under the other to be recycled. These plates can also grind sideways along each other. This theory, called Plate Tectonics, is certainly not fully understood and is continuing to evolve. For instance, recent investigations indicate that the continents may have deep roots that extend through and below the crustal plates.

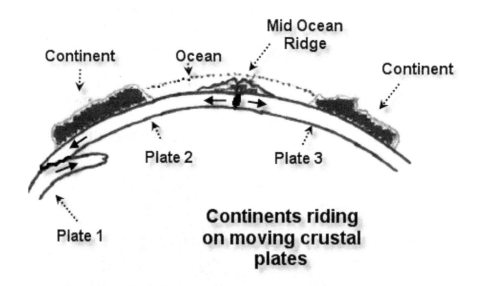

Figure 69 – Plate Tectonics Theory

Was the breakup of Pangaea caused by a major meteor impact? The time frame for the breakup of Pangaea closely matches that of the "Permo-Triassic" extinction, sometimes called "the mother of all extinctions".

In order to explore these investigations with you we have included portions of the "Ocean Floor Map" by Marie Tharp and Bruce Heezen, which were used extensively in trying to understand the events of 10,400 BC. Marie Tharp painstakingly hand painted this map using only millions of electronic ocean floor soundings as reference. Marie and Bruce must have devoted many years in producing this beautifully detailed map for all of us to enjoy. Many thanks!

With the recent mapping of the ocean floor it was anticipated that many questions would be answered. Unfortunately these maps presented more questions than answers. The Oceanic Ridge is an underwater mountain range that is hundreds of miles wide with peaks over two miles high. It winds around the Earth for 35,000 miles. One of the disturbing mysteries is that this Oceanic Ridge has perpendicular fractures and a lack of alignment throughout its entire length. It is conceivable that at one point in time this ridge should have been in alignment. Have major impact events played a role in this erratic formation?

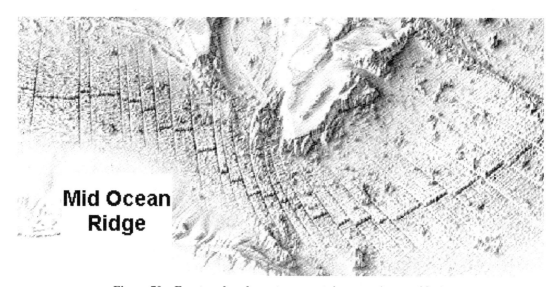

Mid Ocean Ridge

Figure 70 – Fractured underwater mountain range (ocean ridge)

With all this in mind and understanding that The Great Pyramid decisively marks the geo-repositioning of Giza (Africa), we will start from scratch and use this information as a starting point in trying to understand the events of 10,400 BC.

Africa

If the Great Pyramid is showing the geo-repositioning of Giza at 10,400 BC, this repositioning would only be possible under two circumstances, continental movement or polar shift. In this chapter we will look at all the varied information gathered from the continents and try to determine which one of these two solutions is more plausible. Realizing our background is not in science or geology but investigation of facts, this has been a good thing in a way as we had no predetermined answers. You must understand that the purpose of this book is to share opinions and observations.

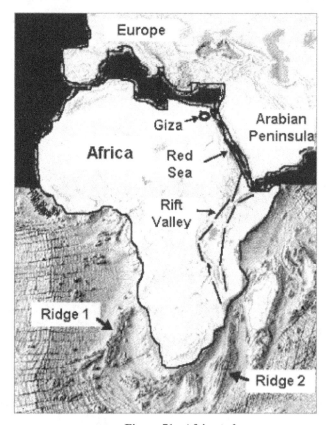

Figure 71 - Africa today

An important fact is the location of the Queen's Chamber within The Great Pyramid. The Queen's Chamber was precisely located on the centerline within the pyramid and this was done to mark Giza's position in the Northern Hemisphere before impacts (Giza's start position). We must therefore try and understand the significance of the King's Chamber being located offset from the pyramids centerline. This positioning was used to mark some type of offset change as related to Giza. Was the The Great Pyramid representing movement of the entire continent of Africa? As Africa was shifting northward by 14 degrees of latitude it was also shifting westward, offsetting Giza to a new longitude.

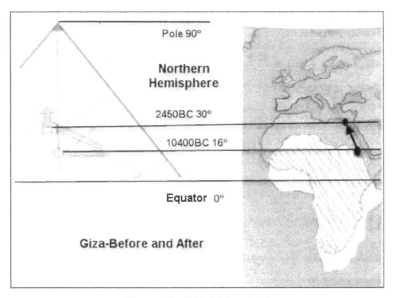

Figure 72 – Did Africa shift?

Testing this theory, if you physically cut out the continent of Africa from an atlas and used it to retrace Giza's movement you would see that it closely follows the path of what is now known as the Red Sea. A line drawn through the Kings and Queens Chambers is oriented much the same as the Red Sea is oriented.

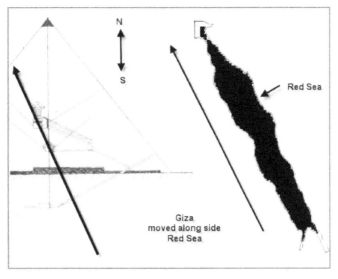

Figure 73 – Red Sea correlation

This indicated that possibly Africa's northward movement had been restricted as it ground along the Arabian Peninsula, causing the formation of the Red Sea and affecting Giza's longitudinal position. One of the first obvious visual points is the long, thin and straight formation of this sea. Less known and not visible is the existence of a very deep trench that runs the entire length of the Red Sea. The floor of this long thin "crack" shows signs of recent lava flows and contains active volcanic pits over 6,000 feet deep.

In fact the Red Sea lies above a major fracture in the Earth's crust which marks the dividing line between two giant crustal plates. Is this evidence that it could be possible that Africa had shifted northward suddenly and not over many millions of years as theorized by science? Also curious is the positioning of the Oceanic Ridge, an underwater mountain range, which travels northward under the Indian Ocean and then suddenly veers west and disappears into the Red Sea. It is not hard to imagine that it was crushed and buried as Africa and the Arabian Peninsula came together.

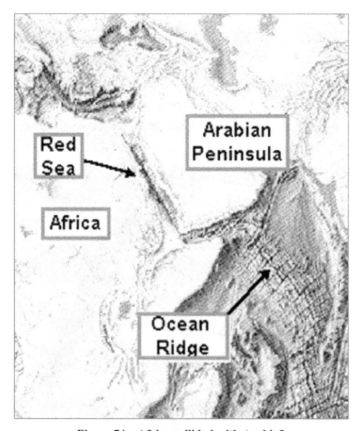

Figure 74 – Africa collided with Arabia?

Investigating the possibility of Africa's movement further and realizing this movement would have been stopped in part by the Arabian Peninsula, you would expect to see some type of proof. Incredibly at this position we find the beginning of the Great African Rift Valley (see figure 71). Science recognizes that as we speak today the entire lower eastern coast of the African continent is being severed. The line of this severance is the Great African Rift Valley, which is measurably lowering itself as Africa separates. Science is aware that this valley will eventually flood with seawater and cause the creation of a new island. What science presently struggles to understand is the cause of this separation because contrary to the norm for this type of movement, no fracture zone is found in the underlying crustal plate that can fully explain this rift. When observing the diagram of Africa you will see that the southern tip of the Arabian Peninsula would affect this southern portion of Africa like a wedge (if Africa had been forced northward) thereby causing the land separation which is creating this Rift Valley.

Upon further investigation (Africa's movement 12,400 years ago) there began to be evidence to the contrary. Indeed, if Africa had been originally 1,000 miles south of its present position, this would likely mean that the Mediterranean Sea would previously have been an ocean (part of the Atlantic Ocean). In direct conflict are the two-mile thick salt fields that form the basin of the Mediterranean Sea. Science feels that for this amount of salt to have been laid down, the sea must have gone through drying out periods (as many as twenty times) and that for this to have occurred, the gap between Africa and Europe (Gap of Gibraltar) must have repeatedly opened and closed over time. The evidence suggests this occurrence took place around 5 million years ago and therefore it seems Africa and Europe have been situated close together for a very long time.

Try to imagine this empty basin, in places two miles below sea level, kept dry by only a sliver of land that held back the Atlantic Ocean and then visualize the catastrophic floods that may have occurred if this land dam gave way suddenly. This would have been a visual sensation almost impossible to imagine. To further understand that the Mediterranean did indeed dry up is to understand the discovery of an existing canyon (not unlike the Grand Canyon), which today is completely filled in with sediments and lies buried beneath the current Nile River. This canyon was formed 5 million years ago when surface waters from Africa reformed the edge of the continent while rushing into this deep basin. This ancient canyon was only discovered by accident when it caused problems during the building of the Aswan Dam in Egypt. How many more incredible finds are set to be discovered below the Earth's surface?

This information did strongly indicate that Pole Shift was a more reasonable conclusion than continental movement (Africa). Pole Shift could well account for Giza's previous position at 16 degrees N. Latitude.

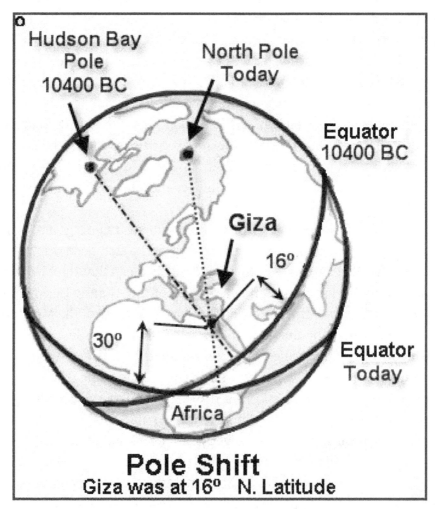

Figure 75 – Did Pole Shift occur – 10,500 BC?

Lastly from Africa is the magnificent Luxor Temple, which was built by the ancient Egyptians in a unique offset construction. Is this another marker that the Ancients were well aware of an Earth Axis re-alignment? There is no widely accepted explanation for this unique construction.

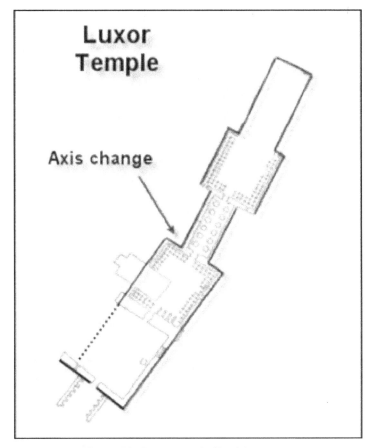

Figure 76 – Luxor was built with two axes

Hopefully at this point you see that this theory is becoming more credible and intriguing. It is now time to examine evidence from all four corners of this planet.

North America and Asia

Something that has puzzled many from the beginning is the position of the "Ice Age" ice fields. Drawing a line directly through today's North Pole we find the positioning of previous ice fields to be solely on one side of this line. On the other side of this line we find little evidence of ice cover, in fact, as stated earlier, it appears millions of mammals existed in what must have been a Temperate Zone. This would be like living in Florida and having Antarctica as your neighbor.

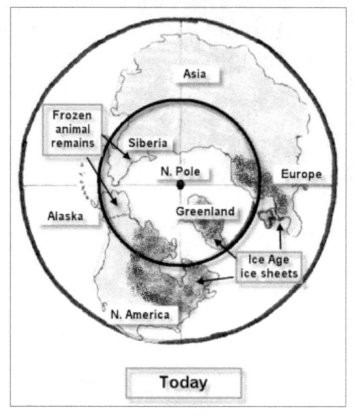

Figure 77– Current North Pole configuration

The established theory is that much of North America and Northern Europe was covered in vast ice fields prior to 10,400 BC; areas that today do not support ice sheet formation. We also know that very substantial amounts of frozen animal remains that relate to this time period have been found in Siberia and Alaska, arctic land areas that do not and could not support these plant-eating animals today. Common sense permits us to assume that the land areas formerly covered by ice must have been originally located within the Arctic Circle and the land areas where frozen animal remains have been found must have been originally farther south where plant growth was possible.

Knowing that frozen remains were found in both Siberia and Alaska, any viable solution would include these areas originally being situated at a more southern latitude. Reviewing an ocean floor map we can see that Alaska is actually part of the Asian continent. Both Alaska and Siberia are of the same landmass (the land between them is currently below sea level leaving the impression of separation).

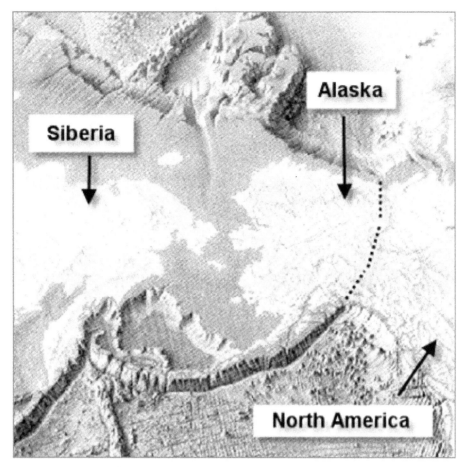

Figure 78 – Alaska and Siberia are one landmass

Again, perhaps another indication that continental movement could explain all anomalies. We find all criteria are met with two straightforward continental movements. Firstly, rotating Asia clockwise from its current position (using a central pivot point) would move northern Europe to a position nearer the pole where its proposed ice fields of the ice ages could exist. This rotation would also cause Siberia (and Alaska) to move in a southerly direction, to a position where millions of Mammoths and other mammals could have lived comfortably.

Our second land movement is of North America. Using the same theory as we did with northern Europe, we must move North America to a position where its ice fields of the ice ages could exist. We must move the North American continent northward into the Arctic Circle.

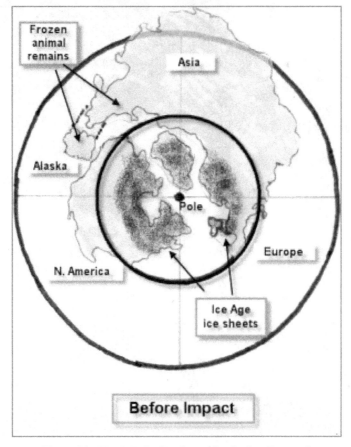

Figure 79– North Pole Configuration before Impact?

With this repositioning of North America and Asia we see a normal sized ice cap centered at the pole. This scenario also could explain the frozen remains of Alaska and Siberia; these landmasses would have been positioned outside the Arctic Circle. These two land movements appeared to solve many unanswered questions and mysteries. Another way to try to understand this would be to imagine Antarctica of today suddenly cracking into three separate land masses representing N. America, Europe and Greenland. If two of these landmasses shifted separately towards the equator then the ice sheets they carried with them would quickly melt (N. America and Europe). If the third land mass moved but remained within the polar region then its ice sheets would be preserved (Greenland).

While evidence was mounting that these types of continental movements had occurred in Earths past, Pole Shift was beginning to appear the more likely scenario as relates to The Great Pyramid and 10,400 BC.

Exploring the possibility of Pole Shift further using the knowledge that The Great Pyramid is indicating that Giza was previously at 16 degrees N. Latitude, we can more easily determine the previous position of the North Pole. Given Giza's previous latitude, a line of possible pole positions would intersect the main North American ice field of the ice ages through the area known as Hudson Bay. The Great Pyramid is also indicating that a 28-degree change in the position of the stars did occur. In accordance with this scenario, all stars would change their position by 28 degrees (relative to the new pole stars) if a 28-degree Pole Shift had taken place. *Incredibly,* if we draw a line of possible pole positions that are 28 degrees from the current pole, that line does intersect the previous line we just discussed at Hudson Bay. Hudson Bay is 28 degrees from the current pole and if Hudson Bay was the old pole then Giza would have been at 16 degrees N. Latitude. *Has the Great Pyramid spoken?*

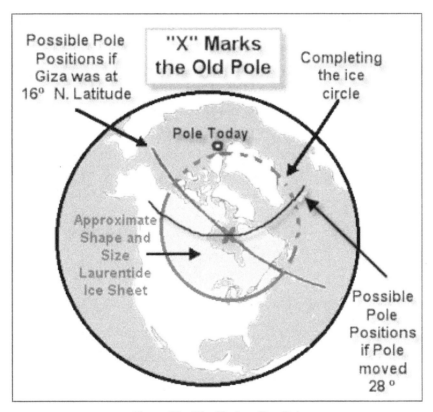

Figure 80 – The Hudson Bay Pole

It would be reasonable to assume that surrounding the old pole there would have been a massive ice cap. We know that during the time referred to as the Ice Ages, the largest concentration of ice fields appear to have been located in the northeastern half of North America (the Laurentide ice field). When observing a map of the Laurentide ice field (see figure 80) we find its land based outer perimeter forms a roughly circular line, which appears to be approaching half a completed circle. Assuming an arctic ice field would be formed from the center out (the North Pole being located near the center), by completing this circular line we find our intersection point in Hudson Bay located near the center of this formed circle. Also we find Greenland, which still holds its Ice Age ice, within this circle as would be expected. These facts are getting hard to ignore.

As for the theorized European "Ice Age" ice fields (which we find are left outside this circle), the evidence indicates that they never really existed. The Finland Impact created much of the evidence that has been interpreted as ice cover and as for any real evidence of ice cover, this comes from glaciers that were formed after the pole shifted closer to Europe, during the 1000 year global cooling that followed this impact event (the Younger Dryas Event). It appears that Pole Shift is a relevant factor and that Hudson Bay was the original position of the North Pole. We must now look at all available evidence with this in mind.

Some intriguing evidence that appeared to indicate that the world had been mapped far back in prehistory are the maps of the ancient "Sea Kings". Shortly after Christopher Columbus's discovery of the Americas, historical maps began to appear that showed N. America and Asia as one continuous landmass.

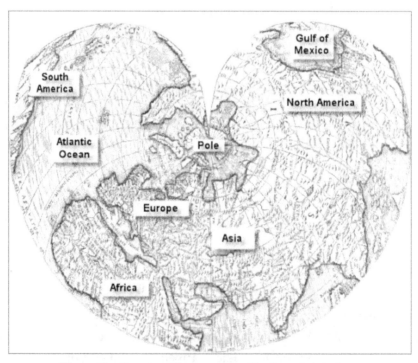

Figure 81 – Oronteus Finaeus map – Northern hemisphere

The Oronteus Finaeus map of 1532 is one of these, although it became famous for a different reason. Not only does it depict North America and Asia joined, it appears to show Antarctica a full 300 years before this land mass was discovered. Not until the 1800's did modern man voyage far enough south to lay eyes on this continent, never mind map it!

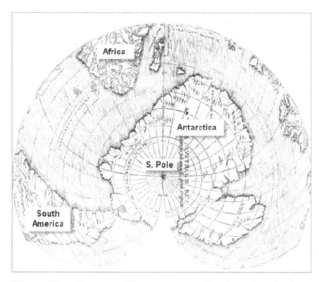

Figure 82 – Oronteus Finaeus map – Southern hemisphere

The important point concerning this map is that it represents land areas that had yet to be explored in the year 1532. The only good explanation (and it was stated as such by some of these mapmakers) is that existing maps of that time were modified using the knowledge they obtained from other very ancient maps. In fact did they have access to maps that were originally completed long before the building of The Great Pyramid? Did a highly advanced civilization completely map the Earth when North America and Asia were one landmass? Science states that the land between Alaska and Siberia was above sea level before 10,000 years ago and this land was only submerged due to the rising sea level that resulted from a melting "Ice Age" ice cap. These continents were indeed joined and Pole Shift could account for the massive sea level rise that in effect separated them.

The Teotihuacan pyramids, which are located just outside Mexico City, are fascinating and important in that they also appear to have been built to mark the stars of Orion's Belt. This ancient site consists of two large structures in alignment and one smaller structure in the offset position, a method of construction and location that closely mirrors Giza. Coincidence? Or is this another memorial marking the events of 10,400 BC?

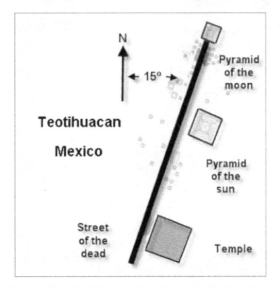

Figure 83 – Teotihuacan is oriented to Hudson Bay

Of great significance, these ruins are laid out along an axis that is positioned 15 degrees off true north. In support of the Pole Shift conclusion, we find the axis at Teotihuacan, if extended, runs through Hudson Bay and this adds further corroboration to the true positioning of the previous pole. Was this site originally constructed over 12,400 years ago so that its axis pointed directly in reference to the previous pole or was it built more recently as a memorial and this orientation has been maintained as sacred through future generations?

Figure 84 – Teotihuacan Aligned to Hudson Bay

Turning to Europe, was the mountain range referred to as the Alps formed in a quicker manner then what is believed by science? Did catastrophic land movement due to impacts take part in the sudden formation of this and other mountain ranges? Could the extinction of the Dinosaurs have been caused by meteor impact much more catastrophic than even the impacts being discussed here today, in fact were the impacts that ended the dinosaur era so powerful as to cause major continental movement? The visual evidence from the ocean floor and other related clues indicate that at one time (or a number of times) the continents did shift suddenly. The mountain ranges of today could very well be holding information as to the timing and location of catastrophic impacts that have taken place throughout our world's history although this science is still in its infancy. Will the evidence contained within these mountain ranges show that the total formation of these ranges has occurred suddenly at different time stages along with the natural movement, which is still occurring today? The events of 12,400 years ago may well have played a part.

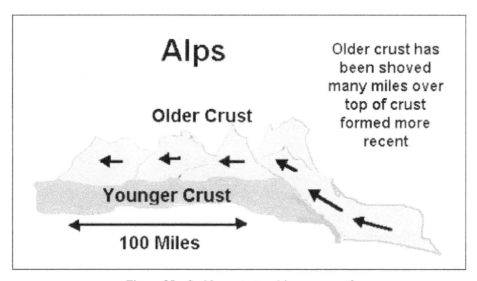

Figure 85 – Sudden catastrophic movement?

There is a very interesting study by Peter Thomson that may help corroborate that some mountain building did occur as a result of the impacts referred by The Great Pyramid. The following diagram represents the Lauterbrunnen Valley formation in the Swiss Alps (you can view a photograph of this valley at www.peter-thomson.co.uk). You will notice that the "before" diagram shows a consistent layering that would indicate this was a stable landmass for an

109

extremely long period of time. When observing the "after" diagram you will notice this same land area has been greatly disturbed. Now the question remains, how and when did this disturbance occur?

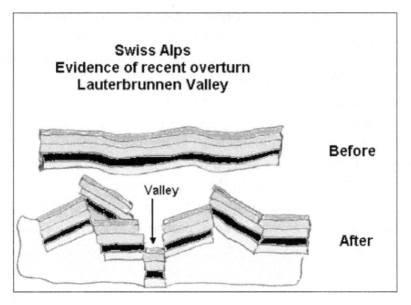

Figure 86 – Valley formation appears to be recent

Peter Thomson's study observes the following points of interest indicating a recent and sudden uplift of this area:

1 – When observing the highest peaks of the valley we notice the sharp edges in this range have not been subjected to long term weathering.

2 – The fan debris at the base of the cliffs is insufficient to what is normally accumulated through cliff face break down over long periods of time.

3 – There is no evidence of glacial movement contained in this valley. Science has indicated that this geographical area was covered with extensive glaciers during the recent past, therefore if this valley was in existence for a long time, you would expect to see scarring left behind from glacial movement. Retreating glaciers of today leave behind definitive evidence of their one-time existence.

This formation and others may be evidence that catastrophic land movement was a result of these impacts 12,400 years ago, in addition to Pole Shift.

South America

To understand what happened in South America we must start with the very relevant fact of the existence of the Lake Titicaca basin, which is located over two miles above current sea level in the heart of the Andes mountain range on the Bolivian border with Peru. As we rise from the floor of this valley, above the level of Lake Titicaca, we see strong geological evidence of the existence of a prior lake level marking. On one side of this basin there is a distinct yellow-white line that runs hundreds of miles. These markings represent the previous shoreline of a massive lake that existed within this basin until roughly 12,000 years ago. This ancient shoreline is tilted and distorted indicating recent and substantial crustal movement has occurred.

Other similar evidence indicating this recent disturbance may have been continental in proportions comes from Charles Darwin's journal. During his investigations of the western coast of South America over a century ago Mr. Darwin references another previous shoreline, this one from the Pacific Ocean. His entries state that running from Bolivia, south through Chile, was a recent looking shoreline that was greatly elevated above the current sea level. Climbing by foot at various locations, his careful observations showed that at Bolivia this previous shoreline was 80 feet above existing sea level and that this shoreline ran continually in an upward manner south through Chile where he last recorded an elevation of 1400 feet above today's sea level. Charles Darwin's initial conclusion was that this amount of movement could not have occurred naturally and he surmised this was probably caused by a cataclysmic event, which forced this movement to occur in a rapid manner.

Let's return to the Lake Titicaca basin; situated in close proximity to and at the same elevation as the old shoreline markings, are the ancient ruins of Tiahuanaco. Some have called these ruins the most ancient ever found yet here we find evidence of a civilization that used levels of technology that baffle the mind. Scattered about are many megalithic stones (weighing as much as 300 tons and as small as 100 tons) that were carved with precision and moved many miles from quarries that have been located. Some have reasoned that these massive stones were used for the construction of portside docking areas. If we were to reassemble these megalithic blocks to their original positions, would we find the continuance of the ancient tilted shoreline that we did previously discuss? Many of these blocks are said to contain the same ancient shoreline markings as is seen close by on the land indicating these stones were part of an ancient port. This would prove that the tilting of the old waterline must have happened after the port's construction, although at this point we should realize that it is very difficult to get reliable and substantiating information concerning these ruins. It is a tragedy that more is not being done to preserve and document these sites.

Figure 87 – Tiahuanaco – Gateway of the Sun

To establish a time frame for this event we must consider all evidence left behind. On ancient stone statues and temples at Tiahuanaco are reportedly found historical carvings of animals known to have gone extinct at the close of the so-called Ice Age. Found so far are carvings of Elephants and Toxodons, species known to have roamed extensively throughout South America until their sudden disappearance about 12,000 years ago. We would have to ask ourselves and be answered this question. How is it possible that a relatively recent civilization made carvings of animals that went extinct many thousands of years earlier? We must consider the likelihood that the origins of this civilization are from 12,000 years ago before the extinction of these animals.

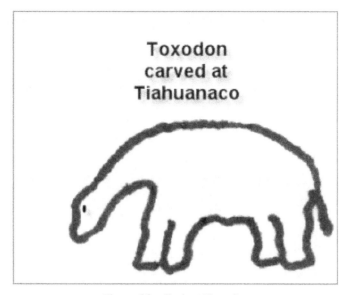

Figure 88 – Extinct Toxodon

Other famous ruins that are found high in the Andes, in or near the Lake Titicaca Basin, are Cusco, Sacsayhuaman and Machu Picchu. Why did these civilizations endure the hardship of establishing their existence so high above existing sea level where food production and survival in general would have been more difficult? Would it be perhaps because of previous experiences? Did the cataclysmic events of 10,400 BC cause mountainous tsunamis and massive destruction, something they were protecting themselves against? Let's look more closely at the Inca creation myth.

Legend has it that the Inca people of South America were created after a great cataclysm by the god Viracocha and that he thereafter "instructed" them. Upon Viracocha's departure over the ocean, he had promised to return. The Aztecs of Central America also had a creator god, Quetzalcoatl, who departed over the ocean and promised to return. When the Spaniards first reached the shores of Central America, numbering only in the hundreds, the fear of battle they felt lay ahead as they encountered thousands of Aztecs was not to be. The Spaniard's arrival was seen as the return of Quetzalcoatl and rather than meeting the anticipated battle, the Spaniards were greeted by thousands of Aztecs in a welcoming manner.

Was the existence of Viracocha and Quetzalcoatl a historical fact? Could they have been survivors of a highly advanced society, possibly charting this New World or just finding themselves lost? Either way they are portrayed as being teachers who left behind great knowledge and Viracocha is credited with establishing the city of Tiahuanaco that we spoke of earlier. An interesting point as to the layout (orientation) of the temple and pyramid located at Tiahuanaco is their difference in axis positioning. The pyramid is aligned to true north while the temple is aligned 4 degrees west of north and points at Hudson Bay (another coincidence?).

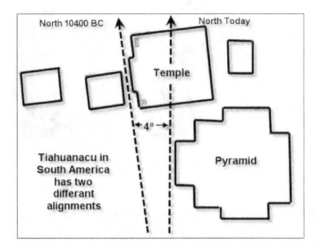

Figure 89 – Why are there two different alignment axis?

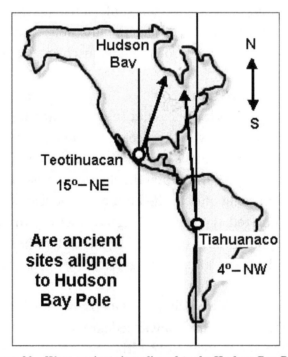

Figure 90 – Were ancient sites aligned to the Hudson Bay Pole?

In keeping with previous discussions, investigation of the ocean floor around South America once again showed indications of sudden continental movement. It is impossible to date this movement, but quite possibly it did occur. It appears that if South America did shift, it moved in a clockwise rotation, which forced the relocation of parts of the oceanic ridge (underwater mountain range) and broke up the southwestern shoreline of the continent. Also, knowing that South America and Africa were once joined and that their coastlines "fit" together, this rotation in reverse would perfectly align these continents to re-mesh.

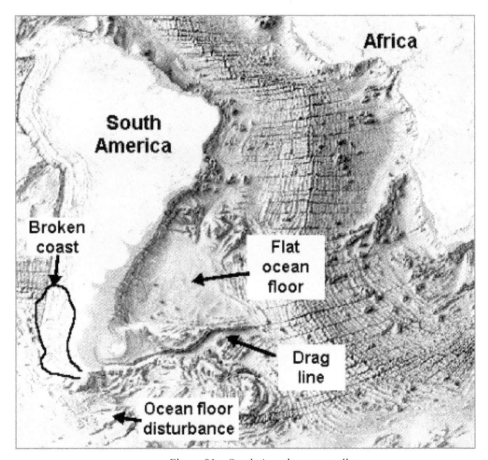

Figure 91 – South American anomalies

It would be reasonable to assume that if sudden continental movements have occurred in the past, they occurred as a result of major impacts. Such impact events are also likely related to continental breakup and collision and major extinction events in our world's history.

India and Indonesia

Science has established that India was once an island in the middle of the Indian Ocean. Science also accepts that India has moved from this original position, in a northerly direction, and collided with Asia (current position). The formation of the Himalayan Mountains, according to science, happened with this collision of India and Asia. Scientists suggest this collision took place slowly and began some 50 million years ago.

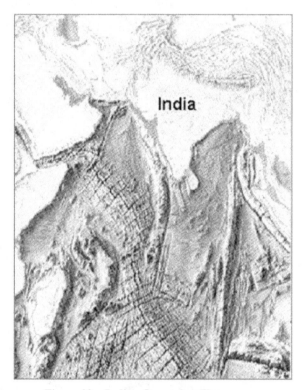

Figure 92 – Indian Ocean Sea Floor

The Himalayas are the highest mountains in the world and rise abruptly like a wall, in a twisted and broken formation, along the entire northern border of India. Close to the Indian border we find the highest mountain in the world, Mount Everest, which holds fossils of sea creatures as high as you climb. It appears this mountain used to be sea floor. Is it possible that India's collision with Asia (and the formation of the Himalayas) took place suddenly as a result of a cataclysmic impact event (the same one that did in the Dinosaurs)?

It is apparent when looking at our map of the ocean floor south of India that lines where etched that scribe India's exact movement. At first glance you see two long ridges that could well have risen as India plowed northward. There are no fault lines under these ridges that could otherwise explain their formation. These two ridges we are studying are very long, reasonably straight, and elevated above the seafloor. If you were to tie a string around your pet rock and drag it in the sand, the ridges produced would show you its exact movement and starting position.

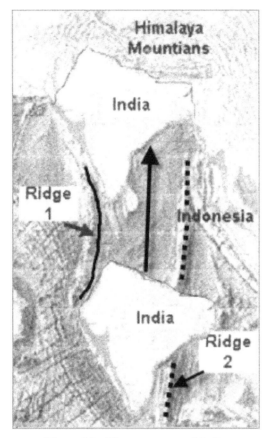

Figure 93 – The movement of India

These facts indicate that these ridges were formed by the independent movement of the island of India through the crustal plate it was a part of. This evidence also then overwhelmingly states that the independent and sudden movement of continents should be considered. If we would consider this a possibility then perhaps many remaining questions would be answered. To understand more accurately the formation of these ridges, consider the following information:

1. India physically did not fit between these ridges. We know, therefore, that the movement of India did not create both ridges.

2. It is obvious when observing the ocean floor that Ridge 1 was created by India's movement. Therefore some other land mass movement must have created Ridge 2.

3. Ridge 2 was formed after India moved north and collided with Asia. If this ridge had existed before then India's movement would have removed it.

4. This second ridge indicates that Indonesia, and the entire continental landmass of which Indonesia is a part, moved north as well. This movement occurred later than that of India (immediately after).

The consistency of these ridges does indicate that the movement of India and Indonesia was sudden and catastrophic. The Himalayas buckled and rose as India slammed into Asia. Science states this movement occurred many millions of years ago as a slow and steady process. Perhaps the reality is that this movement occurred in just one day, the same day that impact "did in" the Dinosaurs. Perhaps science needs to be looking for a related crater larger than the ones being discussed here, a crater much larger than the suspected "Yucatan" crater, a crater over 1000 miles wide, a "Continent Shifter".

There are many anomalies that the theory of Plate Tectonics does not explain. Evidence in rebuttal to the Theory of Plate Tectonics is being further discussed within independent scientific groups. One such group worth investigating is the New Concepts in Global Tectonics Group (NCGT), which began in 1996 as an informal discussion by a group of Earth scientists concerning the inconsistencies of this current theory.

Antarctica and Australia

Now that we are starting to comprehend the amount of force and devastation released by meteor impacts on the Earth, it becomes much easier to understand that Mankind was also dramatically affected by these impacts and our previous history would almost surely have been lost. This fact is corroborated with the use of ancient maps. Maps drawn hundreds of years before Antarctica's documented discovery appear to show this continent positioned at the southern pole. The Oronteus Finaeus map is the best example. Facts we must consider:

1. This map was recorded in 1532
2. Antarctica was not discovered until 1840

3. No known civilization before 1840 recorded Antarctica's existence, never mind mapped it in detail

These facts indicate that Antarctica was charted and mapped prior to the earliest recorded history of Mankind. We know that the Oronteus Finaeus map was drawn using a compilation of numerous other maps, some of which are said to have been ancient. Was the depiction of Antarctica originally drawn over 10,000 years ago and then carefully preserved for thousands of years until it resurfaced in the Oronteus Finaeus map?

A puzzling point for discussion is the depicted size of Antarctica in these ancient maps. Why is it shown as a much larger land mass than what exists today? The communities of modern day explorers trying to solve this mystery feel that there was probably an error made in the reproduction of these maps. Perhaps we should consider the possibility that the error is in our judgment, not the mapmaker's depiction. If we consider this map holds a responsible representation of Antarctica as it existed before the impacts we are discussing, then we must look at all possibilities as to how there could be such a discrepancy in size. Science commonly agrees that Australia was originally part of Antarctica and that they began separating over 50 million years ago, slowly drifting apart to their present positions. Does the Oronteus Finaeus map accurately indicate that Australia and Antarctica were joined as recently as 12,400 years ago and due to the catastrophic impacts being discussed, Australia was severed from Antarctica and forced to move in a northerly direction?

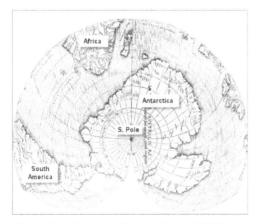

Figure 94 –Orontes Finaeus Map of 1532

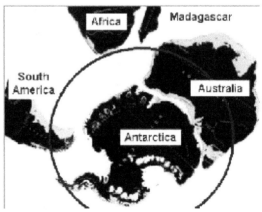

Figure 95 – South Pole 10,400 BC?

Science states that the south eastern corner of Australia and the islands of New Zealand and Tasmania were covered with glacial ice fields up until the end of the so-called "Ice Ages" only 12,000 years ago. If the previous North Pole was positioned at Hudson Bay then the previous South Pole would have been situated much closer to these land areas, making sense of such glacial cover.

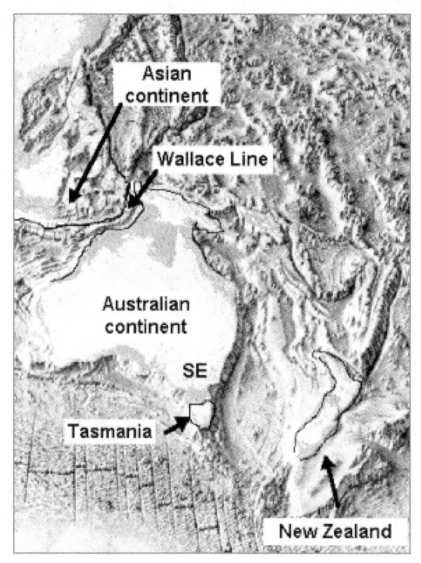

Figure 96 – Did Australia shift recently?

There is another interesting and very important fact to consider. Located where the Australian continental landmass has come together with the Indonesian continental landmass, there exist thousands of islands in close proximity to one another that show two completely different ecosystems. There is an imaginary line, called the Wallace Line, which not only divides these islands but also divides these two very distinctive ecosystems as well. This Wallace Line also closely matches the distinctive line on the ocean floor that marks the dividing line between these two continental land masses. Islands on one side of the Wallace Line have ecosystems matching those found in Australia and islands on the other side (but very close) have vastly different ecosystems matching those found on Asia. The obvious conclusion to be drawn is that these two ecosystems established themselves over perhaps millions of years while these two landmasses were much further apart. Is it possible that these two continental landmasses and their related islands have come together only recently and this is why each ecosystem has so far been able to maintain its individual flora and fauna?

Zep Tepi

So what does this all mean to you? The word Zep Tepi, to the Egyptians, meant the sacred First Time. Could it also have represented a new beginning? The legend of the Phoenix Bird rising from the ashes has its origin in Egyptian myth and seems a fitting representation of Mankind's rise from an unimaginable destruction at 10,400 BC. Substantial evidence strongly supports the theory that the Earth was struck by multiple large meteor-like fragments, possibly from an exploding star (Supernova). These fragments would have traveled through space for tens of thousands of years before encountering our solar system. To more fully understand the consequences, we must try to realize the effects of these mammoth impacts on the ancients and their world.

It appears these impacts did compress large landmasses down in such a severe manner as to displace Earth's core, which consists of lava. This land is still rebounding today (isostatic rebound). This lava displacement caused internal lava tsunamis to race outward from these locations, thereby raising and lowering land masses as far as thousands of miles away and actually moving large land masses and even continents by pushing (shoving) on the lands underneath "roots". Great valleys were formed as land areas gave way and vast regions were raised to lofty heights. Every existing civilization at that time would have been dramatically effected and global ancient folklore still preserves this history today. To further help you understand the true consequences of this event we have included a recent update by one of the co-authors of "Cataclysm".

An Unexplained Arctic Catastrophe - by Derek S. Allen {excert}

As the primary intention of my previous survey of the Siberian-Alaskan 'permafrost' formation was to emphasize its apparent cataclysmic origin around 11,500 years ago, its remarkable geographic extent, the enigmatic character of its composition, and its locally enormous vertical depth, certain other related factors and problems were not perforce touched upon.

The immense depths (4000 feet or more in places) to which the permafrost descends uninterruptedly indicate a vast, but presently invisible, crustal depression (or series of neighboring depressions) subjacent to this formation. Are such features actually collapsed portions of the Earth's crust, analogous to those now represented by the Nansen Basin on the bed of the Barents sea, the collapsed ocean floor off the Kamchatka Peninsula (easternmost Asia), and the Tarfan Depression in Chinese Sinkiang?

These 'collapsed' regions have been explained as a result of crustal shortening resulting apparently from the massive folding-in upon itself of the lithosphere due to sudden withdrawal of material from Earth's lower mantle. Modern man has no experience of such severe crustal dislocations.

The permafrost deposits exhibit every sign of having been swept by water-action into the huge depression (or depressions) they now completely fill. This means that, in at least several places, they filled basins thousands of feet deep and almost certainly scores (if not hundreds) of miles long, again as part of one singular upheaval. The volume of what can be termed 'infill' material is quite obviously colossal, fully justifying the classification of it by Mel'nikov and Grave.

Thus, as seems likely, did a vast tidal wave, or a displaced sea, deposit and mix untold millions of organic remains with boulders, finer detritus, ash and ice all across northern Asia and into western Alaska; and did this deposition, which appears to have been more or less simultaneous throughout the region of permafrost, follow rapidly upon the geophysical creation of the depressions themselves?

Collectively, the organic assemblages preserved in the permafrost deposits constitute promiscuously mixed floras and faunae {plant and animal}, and, most importantly, always in attendance with the bones and teeth of hairy mammoths and woolly rhinoceroses.

As the foregoing data reveal, the kinds of food plants favored by Siberia's mammoths are now known; but, as Howorth wrote over a century ago: 'the food is not now found alive along the Arctic sea, or in the Chukchi land or in New Siberia, even though remnants of mammoths abound all across these frozen lands.

The correct interpretation of these otherwise conflicting details appears to be that former warm latitudes were shifted into polar ones, and that this change involved a realignment of earth's spin axis.

Here is another incredible interpretation of strange phenomena, allow us to introduce an intriguing piece of research by Russian geologist Bekh-Ivanov Dmitri. He has proposed a brand new explanation for the perplexing mystery of the Malta "Cart Ruts".

Malta is an island in the Mediterranean Sea. For as long as records have been kept the appearance of manmade "cart ruts" within the solid rock (limestone) that forms the island has confounded all that have pondered their origin. Many of these very ancient tracks run in pairs as parallel lines and this has led to the "cart rut" designation.

Figure 97 – Malta "cart rut"

Given that science states this rock mass formed millions of years ago, the theory goes that these ruts were formed one of two ways, either they were hand dug in the solid limestone to act as guides for cart wheels or they were worn into the ground by countless thousands of cart trips over the same ground. There are serious problems with both explanations though; some tracks have only one rut, some ruts lead to the ocean and then run out under the water, some cross each other at angles and some merge or diverge. The same rut can vary greatly in width and depth and can also disappear and reappear for no known reason.

Here is the explanation that makes the most sense. Mr. Dmetri begins "I am a professional geologist...In the process of thinking through all the data that I have accumulated for these (many) years I was becoming more and more convinced that the process of mountain building, or at least initialization of mountain building, in different and removed areas as far as thousands of kilometers from each other, has taken place quite recently, only a few thousand years ago. Such wide extent of this process could only point to its worldwide character. Having assumed this, I asked myself a question: if initialization of mountain building has taken place just recently, there should be some traces of it in the history of Mankind. So it happened that my assumption has been tested on the Cart Ruts (of Malta). The results of my research have only strengthened my belief that the chosen way was right."

Malta Cart Ruts by Bekh-Ivanov Dmitri (excerpt)

It is all very simple. Nobody hollowed limestone, nobody used it up and wore it out into dust by wheels. Somebody has crossed fresh dirt once and the trace was ready. Hence, there are all the imperfections of the geometry of the prints. Nobody has worked on them on purpose.

The problem is that my hypothesis assumes that the traces were formed back in a time when the limestone had not hardened yet, i.e. it was in a condition of silt similar to that which covers the bottom of the sea surrounding Malta nowadays. This means that the age of the ruts should be very close to the age of the rock mass. The former obviously should be measured by thousands of years, while the later must be measured by millions of years.

The multi million-year age attributed by geologists to this limestone is not the only thing that contradicts my hypothesis. Even if the limestone was 5 or 10 thousand years old, it would still remain unclear why soft deposits covered hills and plateaus on half of Malta.

It was difficult to believe that tens of geologists during more than a hundred years of geological studying of the islands could be so seriously mistaken. Nevertheless I was confident that the tracks were load-casts of something. So it was necessary to understand in what points the previous researchers were mistaken and made essentially incorrect conclusions.

Paleo Malta was full of life. Giant swans, hedgehogs, turtles and elephants used to live in woods. Herds of pygmy hippopotamus were grazing in swamps. Till now the bone remains of all these animals are periodically found in caves, clefts and old streambeds. All of this data tells us that in recent times the conditions of life on Malta were very favorable for them. Hence, when looking at the modern rocky Malta, it is difficult to imagine where the herds of these animals might have found food. Where are the lakes and swamps in which they lapped? As all above listed animals are not great swimmers, it is obvious that Malta and Gozo were not islands then. There was a connecting bridge, which is now seen as a flat, submarine bar-like elevation at the bottom of the sea.

So, we shall continue. After a catastrophe on Malta and Gozo, many new cliffs were covered with fresh and damp lake silts. Most likely it was then that the bridge connecting Malta and Gozo with Sicily fully or partly plunged under the water. Simultaneously, many areas of the Mediterranean might have risen upwards while others might have plunged. The survivors inhabiting the coasts of reservoirs, people who happened to be in the areas of raisings, suddenly found themselves far from water and surrounded by marsh sludge. It is obvious that they could do nothing else but collect their belongings and leave. Paradise life was over, and struggle for survival began.

In most cases the direction of ruts unequivocally confirms an assumption of resettlement. They go from mountains to lowlands, sometimes lowering on slopes as a serpentine. Sometimes it was impossible to descend straight down to the water because of formed cliffs and then the track goes along them or in an opposite direction, around the cliff, to a place of possible descent.

Under the influence of the wind and the sun they began to dry up. Thus gradually in the process of drying of carbonate silts on surfaces, the crust began to form and thicken, representing not simply dried up carbonate but strong limestone. Due to this phenomenon, tracks stiffened in limestone have reached our time, and we can restore the events that took place some thousands of years ago.

Why did not geologists figure this out before? There are lots of different reasons. Mainly it is because they have confused the cause and the effect. They see cliffs. They see rivers and streams between them. They see loose material carried away by water. Hence they view both mountains and valleys as the result of this process.

Mr. Dmetri's report has been greatly condensed for your convenience. This important new theory can be reviewed in full at *www.cartruts.ru*. It is incredible that these "cart ruts" may be direct evidence of the actions taken in order to survive in the aftermath of this impact event. Further, it appears that the survival of these people was a direct result of this landmass (Malta) being raised above tsunami height. Afterwards, while pushing and pulling their carts (sleds) through the deep muddy mess, they must have relived that terrifying day over and over in their minds. Of course the struggle to survive had just begun, upon reaching the shoreline the realization set in, no more part of the mainland, they were now stranded on a desolate island in the middle of a sea!

Further evidence from the Mediterranean comes to us from the Greek philosopher Plato, as originally reported to him as historical by an Egyptian high priest. The story unfolds that many thousands of years earlier (around 10,000 BC), the Greek army had been assembled to protect their homeland. Then, during one cataclysmic day, the Earth swallowed this entire army. Similar reports of Earth movements are included in ancient myths from around the world.

Now that we are beginning to understand the amount of force released by these impacts, let's try and understand the mechanism by which Pole Shift did occur. It has been 50 years since Charles Hapgood stated that the original positioning of the North Pole was indeed at Hudson Bay and The Great Pyramid is confirmation of his conclusion but that still leaves one question; how did the North Pole move to its current position?

Figure 98 – Pyramids at Giza "Eye witness evidence"

The Great Pyramid is not only showing us that Pole Shift caused a change in the positioning of Giza but incredibly it is also showing the amount of change in position of the Poles themselves. The star shafts are representing a 28-degree change in star positioning that occurred as a direct result of a 28-degree Earth Axis change. Some survivors of 10,400 BC must have fully understood that the 28-degree difference between the old Pole Star and the new Pole Star exactly represents this Axis change.

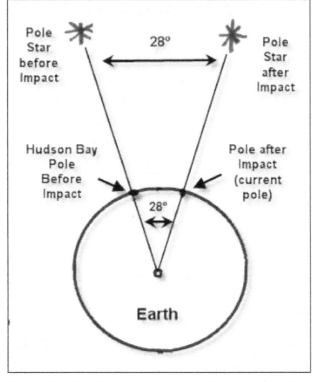

Figure 99 – The "Star Shafts" mark a 28-degree Pole Shift

This Pole Shift did change the position of Earth's equator to its positioning of today. We can now very closely determine the position of the pre-impact equator given the old North Pole was at Hudson Bay. The old equator line would be quite similar to the Ecliptic Line as represented on many of the modern globes of today, only it would be at a slightly greater angle.

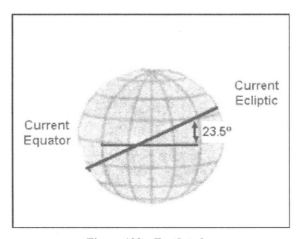

Figure 100 – Earth today

This previous equator line would intersect today's equator line at almost exactly the same angle as the descending passage in The Great Pyramid intersects the pyramid's base line (which represents Earths equator). Have the Egyptians shown us Earth's previous equator with this descending passage? If this pyramid is a "Before and After", then direct representation of the previous equator and previous planetary axis could be expected.

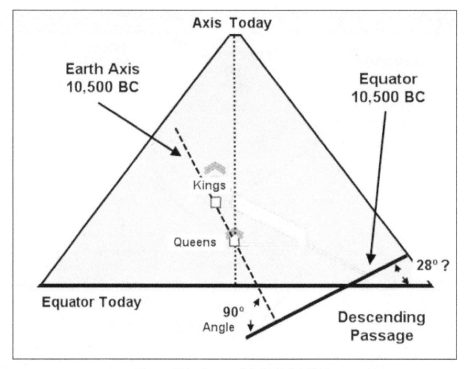

Figure 101 – Internal Axis (Pole) Shift

Other theories exist that this Pole Shift did occur through Earth Crust Displacement (ECD), a violent shifting of the Earth's crust in its entirety around the inner Earth. Under this scenario the inner Earth continues spinning as before although the Earth's crust is physically displaced to a new position (the crust slides around the core). Simply put the Great Pyramid eliminates this scenario. So the question is - how did Pole Shift occur 12,400 years ago?

There is only one other way this Pole Shift could have taken place. Is it possible, as stated in ancient myths, that the earth actually stopped rotating (ESR)? Myths on one side of the planet talk about prolonged sunlight (many days) while prolonged darkness gripped the other side. Did these impacts gradually slow the Earth's rotation to a stop until "forces" unknown (perhaps continuous impacts) caused rotation to start up again only this time on a new axis? Understanding that rotational slowing would not affect gravity, this scenario would be entirely survivable so long as it was gradual like a car slowing from highway speed. Blackboard science would be quick to show you mathematical equations that this scenario is impossible due to these impacts not supplying sufficient force to slow this planet but they are just being close-minded in believing they have all the answers. For example, is it possible that these impacts were accompanied by the close encounter of an even larger sized body (even planet sized) and its gravity supplied the force needed to stop this planet (does the middle pyramid at Giza represent this object passing between the Earth and the Moon?). There are many other possibilities as well; the only certainty is that science does not have all the answers.

This Pole Shift caused the geo-repositioning of every point of land on Earth in relation to the poles. Some temperate land areas (and the millions of creatures that lived there) would suddenly be located in an arctic region, as was the case in Siberia and Alaska. In North America this Pole Shift caused extreme melting of gigantic ice fields that were suddenly repositioned outside the new Arctic Circle. This resulted in a 300-foot increase in ocean levels meaning that any existing coastal civilizations or history of their existence would have been erased forever. Or were they? You decide! Science would prefer to ignore the underwater ruins at Yonaguni (Japan) or simply write them off as a natural formation.

Figure 102 – Yonaguni Japan

A growing number of intriguing underwater sites are being discovered as high technology becomes increasingly available. One such site of great interest, which was accidentally located by NASA satellite photography, appears to be a submerged ancient bridge between India and Sri-Lanka. Ancient Indian myths speak of such a bridge. These NASA photos are available on the Web.

Given that today's oceans are travelling near 1000 miles an hour at the equator due to Earth's rotation, it is reasonable to assume any slowing of the planet could partially displace these oceans from their basins. Also, what would be the effect on the buoyancy of the continents should this planet slow to a stop (no centrifugal force), these continents are only slightly more buoyant than the lava in which they "float"? Is it possible that there was some settling?

Plato, citing even older Egyptian records, clearly infers that meteor impacts and their cataclysmic effects have repeatedly destroyed civilizations.

From "Timaeus"

> There have been, and there will be again, many destructions of mankind arising out of many causes, the greatest have been brought about by the agency of fire and water. There is a story which even you have preserved, that once upon a time Phaethon, the son of Helios, having yoked the steeds in his father's chariot, because he was not able to drive them in the path of his father, burnt up all that was upon the earth, and was himself destroyed by a thunder-bolt. Now, this has the form of a myth, but really signifies a declination of the bodies moving around the earth and in the heavens, and a great conflagration (fire) of things upon the earth recurring at long intervals of time: when this happens, those who live upon the mountains and in dry and lofty places are more liable to destruction than those who dwell by rivers or on the sea-shore; When, on the other hand, the gods purge the earth with a deluge of water, among you herdsmen and shepherds on the mountains are the survivors, whereas those of you who live in cities are carried by the rivers into the sea....it leaves none of you but the unlettered and uneducated, so that you become young as ever, with no knowledge of all that happened in old times.

Plato is just as "current" today as he was over 2,000 years ago; our modern society is still as "young" as ever, having little knowledge of all that happened in old times. The sacred and very ancient writings of Enoch and others refer to a time when the Earth was "knocked over" and was violently shaken. Certainly a pole shifting impact would leave this indelible memory with the Ancients. In fact, the changes that occurred were carefully recorded and then preserved for many thousands of years until an early Dynasty of the Egyptian civilization put in motion an incredibly brilliant plan to mathematically encode this information at Giza in structures that could withstand future cataclysms. The Great Pyramid has so far stood tall for nearly five thousand years.

Exploring further, if the Earth did indeed stop rotating then it is unlikely that it resumed rotation at the same rate as before, the length of a "Day" may well have changed. Is "Time" also represented in The Great Pyramid as "Before and After"? The near doubling in length of the King's chamber over the Queens chamber must have been incorporated for a reason of great importance, could it be that the length of a day near doubled? Ancient myth does support this. For most of our time on this planet we slept twice every 24 hours: now we have an excuse for dozing off at work in the afternoon; it's in our genes.

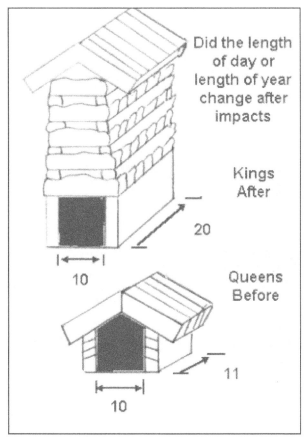

Figure 103 – Do chamber dimensions represent "time"?

The ancient myths also refer to a time when the sun took a new course across their sky, for them this was the beginning of a new "Age" or "Sun". Pole shift definitely would have changed the direction of travel of the sun in their sky.

It appears the evidence left behind by Impacts and Pole Shift has been misinterpreted, which has led science to believe in the existence of what is referred to as the "Ice Age". Pole Shift is a reasonable explanation for the thick polar ice fields that disappeared from North America. Greenland remained in close proximity to the new pole position and therefore still holds its "Ice Age" ice fields. The evidence used to propose thick ice cover in Europe previous to 10,400 BC is based on faulty assumption (Northern Europe was not covered in thick ice sheets while the Pole was at Hudson Bay). On the other side of today's North Pole millions of Mammoths, Wholly Rhinoceros and other creatures lived comfortably in Alaska and Siberia before Impact and Pole Shift caused their demise and subsequent quick freeze.

Figure 104 – Mammoths of Siberia

Previous to 10,400 BC the Earth had normal sized polar ice caps and climate conditions not unlike our own. Just as we see in Antarctica today, the land based polar ice sheets of 12,400 years ago were basically centered on the previous pole positions; there was no ice age just normal ice caps but in different locations. Somehow the Ice Age "snowball" got rolling over one hundred years ago and it will be rolling for some time yet.

How is it possible that this error has been so long lived? Mr. Delair, co-author of "Cataclysm", has stated that part of the problem is research funding that rewards conventional results; results that fit all new evidence into the tired old theories (do not rock the boat). Put another way, *the problem does not rest on the information found but by the way this information is funded.* Therefore those that control the funding are the problem (when a fish stinks it stinks from the head down).

This top down mindset filters down to the street and only a few scientists are willing to risk their career questioning methods and credibility. One such scientist explained it this way "I have read plenty of bogus conclusions or obvious biased interpretations in science journals that are not supported by their own facts. I have read lies. As long as you toe the party line you will not be self-corrected by peers. If you contradict the party line, you will be crushed by your peers without a second look. It is the way it is". Here is one example for your consideration of how these pressures can lead to possible misinterpretation (suppression) of information: One scientist, while investigating the New Siberian Islands in the Arctic Ocean, reported that no sediments dated between 28,000 and 12,000 years ago were found. His explanation, "this may reflect active erosion during that time". Why was the obvious question not pursued – What cataclysm happened 12,000 years ago to erase 16,000 years worth of soil? If cataclysm is not an acceptable conclusion than the next best conclusion will do. *The availability of funding controls the questions (answers)!*

Exploring this further, it is not hard to understand why conspiracy theories are so entrenched in our society; in this case science seems determined on keeping these Pole Shift realities being discussed here buried in the domain of crackpot pseudo-science. Take the case of scientist Michael Brookfield who tried to pursue the idea that the Hudson Bay Arc previously discussed is possibly a giant crater. Unable to get the funding needed to properly investigate, he self-funded an "on the cheap" expedition to Hudson Bay which produced interesting results. He found evidence of multiple concentric rings moving out from this crater as is the norm for large impacts and he reported these findings to the Geological Society of America. That's where it ended, no follow-up funding was ever offered. If there is no conspiracy here, science needs to fund Michael Brookfield as this is serious business. Just in this last year the Chelyabinsk Meteor air burst over Russia injuring 1500 people and multiple behemoth rocks, like the half km wide 2013 Tv 135, had very close encounters with Earth. Incredibly this massive meteor (2013 Tv 135) was not even discovered until three weeks after its close encounter.

Moving on, our previous and present pole positions have been mathematically encoded into The Great Pyramid with astounding accuracy. This was completed thousands of years before the time of the Greek mathematician Pythagoras (500BC), often referred to as the "father of mathematics". Not disputed is that he traveled to Egypt early in his career to study with high priests that were renowned for their wisdom. Was Pythagoras more a gifted "great grandson of mathematics", whose father had lived many thousands of years earlier? This again leaves us with only a much stronger belief in the existence of advanced societies whose true identity still remains an important mystery. Perhaps the stories of lost civilizations are historical fact rather than myth. Perhaps now is the time to stop considering just how intelligent we are and consider the intelligence that we have left behind.

Figure 105 – Message from the Ancients

Chapter 8

The Lost Civilizations

The Mystery of the Sphinx

The image of the serpent was an extremely respected symbol for the Egyptians. The prestigious headdress worn by the Pharaohs was always adorned by the serpent in the form of the cobra (one of the most feared snakes today).

Figure 106– Egyptian (Pharaohs) serpent headdress

A very similar symbol, which was worshiped across the ocean in the Americas, is the Feathered Serpent. Why was a similar symbol so important to civilizations so far distant?

Figure 107 – Feathered Serpent of the Americas

This symbol of the serpent appears to have later evolved into the "Fire Breathing Dragon" symbol that is recognized by all cultures today. Are these all representations of the truly feared meteors that routinely blazed through our atmosphere and brought fire and destruction to many of the ancient civilizations? Perhaps Dragons still do exist. Perhaps the return of the Dragon is inevitable. Now that we are beginning to understand this "message" we must try to understand just who the ancients were that have left it for us!

If we are going to be honest in our quest for the real answers concerning Mankind's past, we must be willing to discuss all possibilities and not place so much emphasis on established theories. It is time to rethink the origins of the Egyptian civilization. A compelling statement by John Anthony West reads as such "Every aspect of Egyptian knowledge seems to have been complete at the very beginning. Egyptian civilization was not a development, it was a legacy". In order to disallow the existence of even more ancient civilizations, we would have to accept the fact that the Egyptians invented civilization and within just 500 years created pyramids with such precise engineering that they are still puzzling us to this day. How is it possible for such a newly formed society to achieve these engineering complexities within such a short period of time? Simple answer, it is not possible, knowledge was handed down from earlier times.

Let us start this new investigation by discussing the characteristics of the Sphinx. A feature that has puzzled many is the head, which looks to have been reconstructed. Understanding that stone monuments built by the Egyptians were always constructed in a precise manner, why is the existing head of the Sphinx so disproportionately small and just does not fit this structure? Another point of interest that arouses curiosity is why does the body of the Sphinx appear to be so much more weathered than the head, the facial features of the head being still very evident whereas the body has been worn almost beyond recognition?

Figure 108 – Sphinx before excavation

Common sense tells us that the Sphinx and its existing head were carved at two different times, the original head being carved many thousands of years earlier than the Egyptian civilization we know and over millennia it was severely weather damaged along with the body. The head was then re-carved by the Egyptians. Egyptology is confusing about these points, they state the construction of the Sphinx was at a similar time as the pyramids (2500 BC) and that repairs to the body took place shortly thereafter (the paws and parts of the lower body have been refaced). But then, why would the Sphinx's body need a great deal of repair in short order while the head seems hardly weathered today? If Egyptology was correct that parts of the badly weathered body were refaced with new stone just a few hundred years after the Sphinx's construction (to rejuvenate and preserve the appearance), why is it these newly positioned repair stones do not show similar wear after thousands of years?

If it could be proven that the Sphinx was in existence thousands of years earlier than the construction of the pyramids, this would confirm the existence of a more ancient advanced civilization (One that Egyptologists wish to ignore). This is rhetoric of course; John Anthony West has long proved this point!

Another part of the puzzle that needs explaining is the sudden appearance of advanced hieroglyphics with little evidence of a prior, more primitive form of writing. As one Egyptologist (Wenke) points out "hieroglyphic writing first appeared in such a developed form that we cannot see the full transition from what was probably pictograph writing". It would be reasonable to assume that all forms of communication are developed from a simple state to a more complex form. It appears we are missing the development stage of these hieroglyphics, yet science states this was one of Mankind's first attempts at written communication. Is it not possible that the development stage took place elsewhere, at an earlier time period? Earlier communication may well have been possible with the use and understanding of written symbols. There have been multiple examples of this form of communication being discovered that have yet to be deciphered, some of which may date to long before the Egyptian civilization. Here is a small sample of one of these findings, called "Old European", which is thought to predate hieroglyphics by thousands of years.

Figure 109 – "Old European"

Just by looking at these signs, this is a definite way of communicating. If you were to compare English printing and Chinese printing today, they would appear to have no similarities at all. Our inability to decipher these ancient symbols today leaves the question, just how advanced were these forms of communication? Perhaps one small symbol speaks paragraphs. We must concede, if indeed advanced civilizations did exist far back in the mists of prehistory, they would have had a sophisticated form of written language. Is it not possible an earlier culture is the true origin of the Egyptians and their hieroglyphics? Did the horrendous impact event of 3,000 BC (Indian Ocean) force the migration of a previous advanced civilization? Was this relocation the beginning of the Egyptian civilization and with them came their knowledge?

Two thousand years ago, on a dark and stormy night, tragedy sent an ancient Greek flagship to the cold bottom of the Mediterranean Sea. There lay a mysterious device for millennia until recent discovery of the shipwreck returned this device to daylight. The complexity and intricacy of this device, called an Antikythera, astounded all that endeavored to understand it.

Figure 110 – Antikythera device

Eventually, painstaking examination and replication by the Antikythera Mechanism Research Project revealed it to be an instrument designed with astronomical and geographical intent. Its many precise gears and levers were used to measure and predict star positions. What so surprised researchers was the level of technology and star knowledge that was built into the device, something more than the ancient Greeks have been credited with. How did this instrument come to be in the possession of these Greek mariners? If one of its primary purposes was determining Earth positioning using the stars, something that would be essential for any global seafaring nation, is it not possible this technology was handed down to the Greeks from the Egyptians? More to the point, was this device handed down to the Egyptians?

There is one piece of evidence that just cannot be ignored, that being the location of the Queen's chamber deep within The Great Pyramid of Giza. This Queen's chamber was built in a precise location to represent Giza's exact position on Earth (16 degrees N. Latitude) before the catastrophic impacts of 10,400 BC. This points to the existence of an extremely advanced civilization, previous to 10,400 BC, which had mathematically calculated Giza's original position (there is no way the Egyptians could have determined this). To understand the positioning of Giza at that time, the lost ancients must have understood that the Earth was round and a great deal more (something modern man has only recently understood). If you openly look at the information contained in this book you will see that this advanced civilization did exist and the level of their intelligence would have been incomprehensible.

Through mythology, for thousands of years, there has existed a legend of a previous island nation containing great wealth, pleasurable lifestyle and a highly intelligent people equal to none. This island sanctuary was reportedly destroyed by a cataclysmic event around 12,000 years ago. Plato's most famous story is of this island. "ATLANTIS REALLY DID EXIST"

Timaeus

To understand the existence of Atlantis is to understand the "works" of Plato. This well-respected philosopher of the classic Greek civilization wrote as follows.

From "Timaeus"

for these histories tell of a mighty power which was aggressing wantonly against the whole of Europe and Asia, and to which your city (Greeks) put an end. This power came forth out of the Atlantic Ocean, for in those days the Atlantic was navigable; and there was an island situated in front of the straits which you call the Columns of Heracles: the island was larger than Libya and Asia put together, and was the way to other islands, and from the islands you might pass through the whole of the opposite continent which surrounded the true ocean; for this sea (Mediterranean Sea) which is within the Straits of Heracles is only a harbor, having a narrow entrance, but that other is a real sea, and the surrounding land may be most truly called a continent . Now, in the island of Atlantis there was a great and wonderful empire, which had rule over the whole island and several others, as well as over parts of the (opposite) continent; and, besides these, they subjected the parts of Libya within the Columns of Heracles as far as Egypt, and of Europe as far as Tyrrhenia.

These writings of Plato have strong indications as to not only the positioning of the island of Atlantis but also the location of other developing societies that were under the control of the Atlanteans (not surprisingly, it appears that the actions of war have followed Mankind from its earliest history). The "Columns of Heracles", as referred to by Plato, has been generally accepted to be a representation of the high points of land on each side of the "Gap of Gibraltar", which is the narrow opening to the Atlantic Ocean from the Mediterranean Sea. It would therefore seem likely that Atlantis was located in the Atlantic Ocean and that the land-mass referred to as the "opposite continent" would be North and South America. Plato states the Atlanteans held influence over parts of the opposite continent and controlled the northern portion of Africa up to and including Egypt. As we are aware of the exact geo-positioning of Giza prior to Pole Shift, it would make sense that *it was the Atlanteans who had calculated Giza's previous position*. Also, knowing they had control over parts of both the Americas and Africa, they were surely an advanced seafaring civilization that crossed the oceans at will using the stars to guide them. This catastrophe of 10,400 BC changed their world forever!

From "Timaeus"

But afterward there occurred violent earthquakes and floods, and in a single day and night of rain all your war-like men in a body sunk into the earth, and the island of Atlantis in like manner disappeared, and was sunk beneath the sea. And that is the reason why the sea in those parts is impassable and impenetrable, because there is such a quantity of shallow mud in the way; and this was caused by the subsidence of the island.

The cataclysmic impacts we have been discussing closely agree with the given time of destruction for Atlantis and would have been in keeping with Plato's story of the earth rocking, the land sinking and the oceans moving. It is easy to understand that these impacts could be responsible for Atlantis disappearing beneath the waves forever. Let us now investigate what level of civilization existed in Atlantis. Plato states the following:

- They "brought two streams of water under the earth which then ascended as springs, one of warm water and the other of cold"

- They " had such an amount of wealth as was never before possessed by kings"

- They "dug out of the earth whatever was to be found, mineral as well as metal"

- They "had sufficient maintenance for tame and wild animals"

- They "employed themselves in constructing their temples and palaces and harbors and docks....a marvel to behold"

- They "dug a canal 300 feet in width and 100 feet in depth...making a passage from the sea....leaving an opening sufficient to enable the largest vessels to find ingress....constructing bridges....raised considerable above the water...there was a way underneath for the ships"

- The "canal and the largest of the harbors were full of vessels and merchants coming from all parts"

According to Plato, this society showed very advanced elements of both design and engineering. Mentioned are the many canals coming to a central point in a circular manner, apparently to ease the access for great many trading ships. Atlantis is characterized as a superior seafaring nation who must have understood navigation by the stars. With their apparent knowledge and understanding of star location and Earth positioning, it is reasonable to believe they knew the Earth was round and had mapped every corner.

There must have been survivors from this advanced society of Atlantis and these survivors were very likely responsible for measuring and recording the repositioning of Giza and the Poles. Plato's description is long and detailed and if accurate, it must have originated from those survivors that once called Atlantis home. Possibly the survivors had some advanced warning of what was to come and based on their own ancient teachings they knew best how to survive meteor impact or possibly the survivors were simply in the right place at the right time. Either way, the survivors knew which stars were the previous Pole Stars and they knew Giza's previous latitude, so with fortitude they set themselves to understanding the changes that had occurred. Here is how they determined two of the most important.

1 – The amount of Pole Shift was determined by measuring the angle between the old Pole Star and the new Pole Star (make the "peace sign" with your fingers and hold it up to your eye so as to pretend you are sighting two stars at once – the angle between your fingers is the angle the ancients measured). This angle (28 degrees) has been recorded within The Great Pyramid by the positioning of the star shafts.

2 – Giza's new latitude (current latitude) was determined by measuring the angle the new Pole Star was above the horizon. Both the "before and after" positions of Giza have been recorded with the positioning of the chambers within The Great Pyramid. [Understanding the 14-degree latitudinal relocation of Giza was directly caused by polar relocation of 28 degrees, we know the difference is due to the fact that the pole shifted in an angular motion and not directly in line with Giza]

The reference points contained within The Great Pyramid leave powerful evidence that this information was handed down to the Egyptians from the Atlanteans, two connected societies separated by many thousands of years. Plato helps confirm this when he states that the story of Atlantis originated with the Egyptians.

Visiting Atlantis by Edmund Marriage (Excerpt)

It was the law maker and one of the most respected Greek citizens Solon 638-558 BC, who had gained knowledge of Atlantis from Sonchis and other priests of Neith in the Nile delta – *"O Solon, you Greeks are all young in your minds which hold no store of old belief in a long tradition, no knowledge hoary with age"* - Solon had decided to travel and consult the wisest men in the region, and to write down information from their old history records. It was this manuscript that Plato used in his dialogues Timaeus and Criteas.

The records of Atlantis were passed down in the time honored ancient Egyptian oral tradition, and inscribed on a stele, much later inspected and confirmed by Crantor, when he checked on the Plato account. Later Sais was destroyed during Alexander's Macedonian conquest and the stele was lost, hopefully buried awaiting discovery.

Direct and associated knowledge of Atlantis in the Atlantic appears independently of Plato in pictures, maps, written and oral records, from all over of the world. *The deeds of kings on the Egyptian King lists going back 36,525 years*, match the known progress of Homo Sapiens, and describe historical events which help fit together and make sense of the story of our ancestors. Thoth the inventor of writing was said to have ruled on an island located in the west. Atlantean survivors are found where they would be expected to have found dry land on the Atlantic coasts.

It appears that the Egyptians inherited all-important information of that time. One great unsolved mystery is the possible existence of the fabled "Hall of Records", which is rumored to be contained within Giza. All of the answers to these important questions we have been discussing in this book may be contained within this "Hall of Records", which very likely did exist at one time. Do these mythical documents still remain within Giza or have they been sealed from Mankind forever?

After the impacts of 10,400 BC the survivors of the advanced civilization known as Atlantis were unconditionally forced to re-establish their society in a new homeland. There is little evidence that Giza (Egypt) was the next homeland for the surviving Atlanteans (and their descendents) although it must have been very sacred ground for them as here we find their incredible masterpiece, the Sphinx (the Sphinx faces directly east and therefore must have been built after Pole Shift). For this reason, their new homeland likely was located no great distance from Egypt. The confusing yet mind-expanding ruins at Baalbek in Lebanon fit the criteria and if we truly wish to discover this interim civilization, this is where we must start.

The Golden Age Project

A very interesting point concerning Baalbek is its location; it appears to have been constructed as a singular temple structure with no apparent ties to any known settled areas (in the middle of nowhere). Yet at Baalbek we find foundation stones realized to be the largest stones in existence that Mankind has ever cut and positioned. Many of the stones used in this foundation exceed 300 tones and even more remarkable is the existence of three stones weighing nearly 1,000 tones. Incredibly, these three stones were not used as base stones but were raised and positioned upon smaller stones that formed the base below. All of these stones were fitted with such precision that even today there is not the slightest gap between them (How is this possible? Yet to be understood!). Is this concept of size and strength similar to the construction that was used to build the island fortress of Atlantis?

Figure 111 – Balbeek (look for man at left)

These ruins of Baalbek have throughout history been consistently related to Roman society. Perhaps the construction reaching upward from these massive foundation stones could be an example of early Roman society although these foundation stones must have been positioned thousands of years before documentation of the Roman civilization, long before even the Egyptian civilization. When observing the previous picture, there are obviously two different technologies of construction used, each with different weathering patterns. The top is consistent with Roman engineering and shows much less natural weathering suggesting the base is of a much earlier time period. There also is a significant difference in caliber of construction. The base was meticulously engineered and remains true to this day whereas the Roman construction was done in a less precise manner using smaller stones (even so, the Romans "Temple of Jupiter" must have been spectacular). Why did the Romans feel this was an important location for the construction of a temple? Baalbek is certainly a great distance from Rome and not close to any important Roman outposts. Was this temple base in existence when the Romans just happened upon it and they felt this was a sacred marking? Did they feel compelled to add to this sacred site?

Baalbek is a very important clue in understanding the existence and possible location of the lost civilization between the Atlanteans and the Egyptians. Perhaps these foundation stones are a very important factor in understanding the missing, highly advanced society of which the Egyptian civilization was born. It seems likely that this ancient platform would have been built within close proximity of its mother city, which may be the home of this lost intermediate civilization.

There is a very important scientific study taking place today by Edmund Marriage, who is investigating the existence of a lost civilization which he believes was located in an area close in proximity to Baalbek, known as Kharsag. Edmund has kindly contributed this important information to share with our readers.

Proof of an Advanced Civilization by Edmund Marriage (Excerpt)

From Plato's Dialogue The Criteas

"O Solon, you Greeks are all young in your minds, which hold no store of old belief on tradition, no knowledge hoary with age. The reason is this. There have been, and will be hereafter, many and diverse destructions of mankind, the greatest by fire and water... after the usual period of years, the torrents from heaven swept down like a pestilence, leaving only the rude and unlettered among you.

Thus the story current also in your part of the world, that Phaethon, child of the Sun, once harnessed his father chariot, but could not guide it into his fathers course and so burnt up everything on the face of the earth and was himself consumed by a thunderbolt – this legend has the air of a fable, but really signifies a declination of the bodies moving around the Earth and in the heavens. "

Modern science led by astrophysicists and geologists have clarified the accuracy of these statements and identified dates, details and causes of these diverse destructions, including the virtual wipe out of most species at the end of the ice age around 10,500 BC, and the destruction of the Bronze Age city states from the Mediterranean to China centering around the tree ring date of 2,345 BC.

Both these dates are huge turning points in the history of our species, the first requiring a complete restart of agriculture and civilization in Southern Lebanon by a small advanced group of catastrophe survivors, known later as Gods. The second requiring the survivors to re-establish the Golden Age of social organization, originally delivered by these Gods, but which in reality did not widely recover, to our cost.

In the years 1897 and 1898 the University of Pennsylvania conducted an archaeological dig and investigation at the site of the Sumerian Nippur library some 80-km south east of Babylon. More than five thousand years earlier the scribes in Sumer had recorded in cuneiform on clay cylinders and tablets, a copy of the story of Enlil and Ninhursag, which were prized library records. These damaged tablets, which had survived almost intact in the rubble from the time that the library was fired and destroyed, were recovered and taken back to Philadelphia where they were stored in the Museum basement.

Archaic Tablet 8383 - Sumerian Nippur Library

Christian O'Brien in his book The Genius of the Few, first published in 1985 and co-authored by his widow Barbara Joy, sets out the evidence that the story of Enlil and Ninhursag relates to what the Sumerians called Kharsag (head enclosure), which was one and the same as the Hebraic Garden of Eden, and that this record was a pre-historic reality rather than a biblical myth.

O'Brien had become the scholar who continued the work of Samuel Noah Kramer, who was one of the world's leading Assyriologists, and a world renowned expert in Sumerian history and language. O'Brien concluded that the south Rachaiya Basin (in Lebanon) met the requirements as being the most probable location of the Kharsag/Eden site. And further that; a group of culturally and technically advanced people who settled in this inter-montane valley in the Near East had established an agricultural and teaching center at about 8,200 BC. (Now re-calibrated to about 9,300 BC).

He derived his choice for the location of Kharsag from a wide range of disciplines, including the descriptions of the area given by Enoch when he was taken to meet the Great Lord and recorded all that was going on. O'Brien finally used the French surveyed 1:20,000 map of the area, to eliminate three other possible inter-mountain basins before deciding that the Rachaiya south basin site, best matched the evidence.

The cylinders and tablets, recording the Kharsag Epics, form part of the Nippur collection held at the University Museum, Philadelphia in the USA. They describe in detail the agricultural, and advanced technical activities of the primary Sumerian Gods (An, Enlil, Ninhursag and Enki). The details within the Kharsag Epics are supported independently by the Chronicles of Enoch, and the early chapters of Genesis. They provided source material, free from later Accadian and Babylonian corruptions, which have devalued and confused the records.

The Sumerians had no doubts that their Gods were real people who were capable of wonderful feats, led by the single God An. Abraham of Ur and his ancestors would also have been aware of this. The high culture of superb social organization with its teachers and craftsmen, maintained within the Druid traditions, had delivered the Golden Age referred to in cuneiform, the Bible, by Hesiod, Lucretius, Seneca, the philosophers of Alexandria, and many others. Hadrian honored these self-same Gods and their progeny, with the still standing wonder of the Pantheon in Rome.

We should reflect upon their benevolence in the messages they have left for us.

Is it possible the discovery of Kharsag will be a very important step in understanding our past? You can view Edmund's related video presentations, which can be found at www.goldenageproject.org.uk.

Fingerprints of the Gods

The learned men of today explain our civilization history relating back to the time of the ancient Egyptians. Recorded history of Mankind's civilization, as we are aware of it, has encompassed only 5000 years and this has been studied by historians and scientists and many other eager minds trying to truly understand the occurrences of the past. Graham Hancock found and felt there was a lost "Ice Age" civilization reaching back tens of thousands of years and he stated "I have learned to respect those long forgotten and still only hazily identified Newtons and Shakespeares and Einsteins of the last Ice Age".

Many things in the earliest Egyptian society show highly advanced technology and techniques (such as architecture, domestication and agriculture) and this leads to the belief that much of this knowledge was passed down from an advanced civilization previous to the Egyptians, a civilization that the Egyptians were powerfully connected to. One small piece of information that strongly confirms this is the Egyptian's own Kings List, which dates back over 30,000 years from present history. What will be even harder for us to comprehend is the true understanding of how advanced these civilizations were. It is very important to keep our minds open and further the investigation as to the existence of these societies. When the lost civilizations are truly found to have existed, it will be incredible to see the stories of these ancient times unfold.

The evidence of these lost civilizations is scattered all around us and should not be ignored. For some reason this information and these facts do not seem to take any priority with the mainstream educators in discussions concerning the history of ancient man. Is it just easier for these educators to dismiss this evidence or are there other reasons this is not being discussed? In order to open a combination lock you must know all the numbers. Part of the problem we all face in trying to comprehend this situation is that most of the evidence of the ancient societies, those that existed before 10,400 BC, has been destroyed or buried due to global devastation. We still have glimpses of these lost societies though, such as the submerged Bimini Road and Andros Platform of the Bahamas.

Perhaps one day we will have more than just a glimpse. Just recently Google introduced Google Oceans and shortly thereafter reports of strange manmade markings on the ocean floor caused a stir. Google was quick to respond that these markings were sonar echoes that traced the path of the mapping ship but we should never assume such statements are fact; these markings are exactly where Plato reported Atlantis' canals would be. Only time will tell.

Figure 112 – Ocean Floor – Outside the Pillars of Hercules

The survivors that recorded the impacts of 10,400 BC must have re-established their particular society over a period of time. How else is it possible that the information contained within The Great Pyramid was kept in circulation for a period of 8,000 years previous to its construction? Yet no such civilization center has ever been found. Did this interim civilization suffer a similar fate? Were other catastrophic events, such as the impact in the Indian Ocean at 3,000 BC, also responsible for a further loss of evidence? In this case though, evidence was not entirely lost, in fact powerful evidence is staring right at us. The ruins of the Sphinx and Balbeek and other sites have been misinterpreted for years.

One misinterpretation is the cyclopean connection between many of the world's most ancient and mysterious ruins. Separated by oceans we find ruins that incorporate impossibly large megaliths in their construction. One truly remarkable example is the "fortress" ruins of Sacsayhuaman in Bolivia. Here we find hundreds of massive stone blocks (some weighing more than a diesel locomotive) that have been carved, transported and exquisitely fitted together so as to build 18 feet high walls that stretch across the landscape. After investigating the site in 1609, the Spanish explorer Garcilaso de la Vega wrote "It passes the power of imagination to conceive how so many and so great stones could be so accurately fitted together as scarcely to admit the insertion of the point of a knife".

Figure 113- Sacsayhuaman

Another very important site is the ruins at Gobelki Tepe in southeastern Turkey. Klaus Schmidt of the German Archaeological Institute has uncovered three dozen "T" shaped stone megaliths, each ten feet tall with beautiful carvings of animals such as foxes and scorpions and lions. Ground penetrating radar has revealed the likely existence of 250 more megaliths that are yet to be uncovered. What is most intriguing about this site is its preservation and extreme antiquity. A reliable dating process has revealed this site was originally constructed over 11,000 years ago, which is a very important fact. Gobelki Tepe may have been established shortly after the impacts of 12,400 years ago and may well be an open window into our forgotten past.

Figure 114 – Gobelki Tepe

The evidence at Gobelki Tepe also shows that there was extreme care taken in the preservation of this site. The creators of these monuments painstakingly tried to ensure their preservation by carefully covering the entire area with protective dirt and soil, no small task indeed. Incredibly, this same procedure is evident in Peru, across the oceans. The debate over the very ancient ruins, known as Caral, is not how this city started but how it ended. Important monuments and pyramids were carefully protected and buried and then it seems the site was mysteriously abandoned. Coincidence (or not!). This only leads back to the further possibility that the Ancients were globally connected and that much more evidence is yet to be discovered.

Message from the Ancients

Press Release - Cairo: The entrance to Egypt's cave underworld has been sealed shut just two years after its modern day discovery. Access to the tomb leading to these natural caverns, located beneath the famous Pyramids of Giza, is now blocked by a metal gate set in concrete. This came in the wake of revelations in the press last summer that British explorer and writer Andrew Collins had in March 2008 located a previously unknown opening into a natural cave system long thought to exist at Giza, but never before explored in modern times....

Despite these recent discoveries, Dr Hawass has publicly denied that any natural cave system extends from the tomb, stating that what exists beneath the ground are catacombs carved by human hands, something that Mr Collins disagrees with strongly. "We have dozens of clear photos, along with film footage, that make it clear that extending from the Tomb of the Birds is an extensive series of cave passages that almost certainly reach beneath the main pyramid field," Collins said....

Collins's evidence is supported by the memoirs of explorer Henry Salt who in 1817 gained access to the same cave system, and explored them for a distance of "several hundred yards" before coming upon four spacious chambers, from which went various labyrinthine passages.

Perhaps this cave system is why Giza is the center of the greatest story never told? Did it provide sanctuary for a band of survivors from a pre-impact super civilization? Were they drawn to this spot originally because all these caves would be a natural habitat and a safe place against attack from other cultures and were they drawn back to them when they saw those meteors coming in the night, that impact coming right at them for weeks in advance? For sure they would again find safety in these caves. And if at that point there were these impacts and their home mainland disappeared and everything they knew was lost, this spot would certainly become a sacred and holy place. So if their descendants were going to build a memorial to thank the gods this would be a good place to do it. It would be the same today if something happened and your little piece of land was protection for you, it would be a sacred place and anybody would have a hard time getting you to abandon it.

They couldn't put bars around the Great Pyramid though; we now know there was a pole shifting impact 10,400BC and incredibly the Great Pyramid even specifies that impact was on the previous North Pole at Hudson Bay (it is the Queens Chamber that represents "before this impact" that is on the pyramid's center line). It is no surprise we find a 500 km wide bullet hole crater at this location. If we're talking about an impact of this nature, if you have an impact that doesn't leave the regular geologic trail, doesn't leave that crater dish then the rock was so big and so fast and the mass of the material was so great that all the direct force was straight down until it got to the level of molten lava and it's really at that point it doesn't come back, it doesn't spill out, it just keeps going and major turmoil happens under the continent.

Let's discuss the Great Pyramid one last time. The capability of the designers, the execution of the project and the quality of the workmanship would be very difficult for us to duplicate today. The true understanding of planetary geometry and global positioning shows us extremely advanced skills. There would have been an abundance of planning and a tremendous amount of detailing in order to construct this monument and this would have been completed long before the cutting of the first corner stone. Why was such an architectural phenomenon constructed with such precision and effort? Why entomb within it historical knowledge of the distant past, knowledge from the survivors of an advanced civilization that existed 8,000 years previous? Was The Great Pyramid built as a memorial in honor of the lost ancients or as a record of history for future generations? Given the indestructible nature of this monument we can safely assume it was constructed for both past and present.

We feel the ancients have left a decisive message in this structure and we have attempted to show that the information put forward in this book has been verified by enough accepted authors out in this field. So why is science so steadfastly against the truth? Perhaps because the real message the ancients are leaving doesn't happen to be the one the establishment figured out in time so the "Ice Age" is going to be their convenient reason for everything or perhaps they know there was a polar shift but it doesn't fit their plan. But science can't just keep getting valid information supporting the other side of the argument and continually just dismiss it but for some reason in their society if you're a no name, you're a no name. Information is information though and you can't really dispute information, it doesn't really matter who it comes from and the truth will always survive anyway. The trouble is the cat is out of the bag now so the more people that care about this type of thing the more proper information science will have to deliver but the powers that be know there is so much BS in this world and people are so busy trying to live that who the heck has time to think about this kind of stuff. So they will continue to feed it to us as much as they can and most will swallow it and move on.

This situation is perhaps akin to the Wall Street collapse of 2008 where other big institutions said trust us. In that case as well many individuals stood up to sound the alarm but were ignored and swept aside by the 'powers' until eventually a world financial meltdown exposed the fraud. If the next meltdown is some mountainous space rock striking Earth we could find ourselves searching for grubs in the ground rather than diamonds. So what can we do on our own? Think for yourself and do not always believe what we are told. Have enough interest in the big picture to understand the problems and basic facts so as to make your own conclusions. Don't feed from the information trough; choose your own meal. The only way we have a chance at becoming an honest society is by making people accountable.

Some of you may remember a book called "1984" which portrayed a future of being manipulated by Big Brother. This book was considered fiction at the time yet the possibility of our civilization reaching that point frightened many. That same fear has returned. Our leaders of today prefer to keep us in the dark and the main stream media operates within a predetermined box. While researching this book and trying to find information available it has become pretty obvious that any information that challenges the throne is considered completely foolish. One of the most frustrating things has been learning just how much we are controlled, we are being spoon-fed to be a certain society the way they want it to be. Unfortunately the only way this type of thing can ever be changed is with the younger generations understanding that we are being pushed around like a bunch of little puppets and the only way to handle these situations is to stand up as a group and say this is not fair, this is not something we like. The situation is so easy to fix its craziness, if we as people had one thing in common that we all lived in a harmonious type of situation it certainly could be achieved but of course that's too easy and it will never happen because that's the attitude we've been taught. We are not allowed to make decisions for ourselves and the government doesn't listen to us anymore. Our elected officials do not consider us a priority; it seems the question is - Who is their priority?

The ancients worked together on the Great Pyramid for no selfish reason; it was simply an indestructible message for the future. Will we listen? Up until their time, right up until our time, there has been little protection from incoming space debris but now for the very first time we have the potential to defend ourselves. Couldn't some of this money in the world's defense budgets be diverted to protect us from cosmic threats? The world in general should work in unity for the one purpose of protecting ourselves because a major impact is definitely going to affect us all so perhaps some of our priorities should shift a little bit upwards. In our civilization we are capable of understanding the messages that have been sent to us from the Great Pyramid, whether or not we decide to use them and pay any attention to them is up to us.

We have been given a glimpse of the pre-flood world where a young man setting out on adventure would encounter a kaleidoscope of strange and wondrous beasts. He could have watched some 20 species of Elephants that have gone extinct; he could have had the pleasure to watch Mammoths graze by the thousands. It was an age of advanced ocean going ships and accurate world maps; it was a Golden Age. *We now can begin to understand this message from Giza and if we truly have a chance to understand the ancient societies, we will have a better chance to truly understand ourselves.*

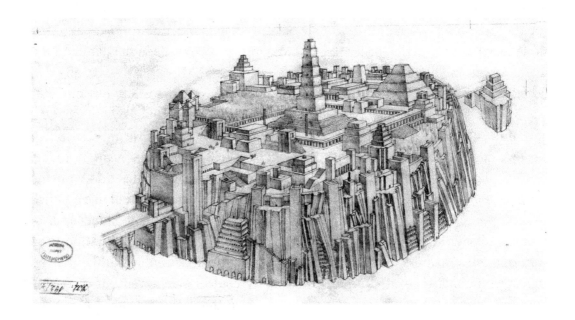

CPSIA information can be obtained
at www.ICGtesting.com
Printed in the USA
BVHW020451200123
656609BV00002B/120

9 780981 128115